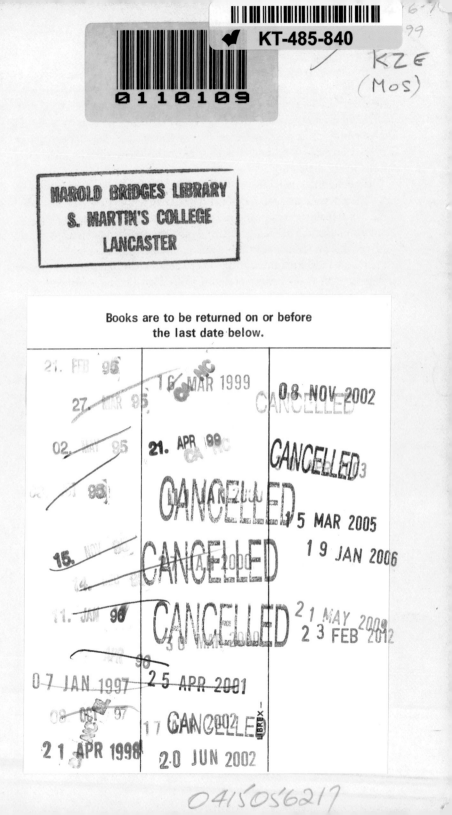

Gender planning and development

Gender planning is a new tradition whose goal is to ensure that women, through empowering themselves, achieve equality and equity with men in developing societies. *Gender Planning and Development* focuses on the interrelationship between gender and development, the formulation of gender policy and the implementation of gender planning practice. Its importance relates to the fact that current development policy, because of incorrect assumptions, often, if inadvertently, discriminates against or 'misses' women, while even correctly formulated policy too often fails to get translated into practice.

Recent feminist theories and current debates on women, gender and development provide the conceptual rationale for key principles of gender planning. These relate to gender roles and needs, to control over resources and decision-making within the household, and to Third World policy approaches to women in development. From extensive research and teaching experience in gender planning, Caroline Moser shows how such principles are translated into methodological procedures, tools and techniques that are integrated into a gender planning process.

She considers whether constraints in the implementation of gender planning are technical or political in nature, and analyses both institutional structures and operational procedures to integrate gender particularly into the project planning cycle. The role that training plays in creating gender awareness and providing appropriate tools and techniques is emphasized with practical exercises for trainers provided. Recognizing the importance of women's organizations to confront subordination, Moser highlights the entry points for such organizations to negotiate for women's needs at household, civil society, the state and global levels.

Gender Planning and Development is a unique introduction to Third World gender policy and planning issues. It will be essential reading for academics, practitioners and trainers in the field of development, and for students of anthropology, development studies, women's studies and social policy.

Caroline O.N. Moser is currently Senior Urban Social Policy Specialist at the World Bank, Washington, DC.

Gender planning and development

Theory, practice and training

Caroline O.N. Moser

Routledge

London and New York

First published 1993
by Routledge
11 New Fetter Lane, London EC4P 4EE

Simultaneously published in the USA and Canada
by Routledge
29 West 35th Street, New York, NY 10001

Reprinted 1994

Typeset in Times by Michael Mepham, Frome, Somerset
Printed and bound in Great Britain by
Mackays of Chatham plc, Chatham, Kent

British Library Cataloguing in Publication Data
A catalogue record for this book is available from the British Library

Library of Congress Cataloging in Publication Data
A catalog record for this book is available from the Library of Congress

ISBN 0-415-05620-9 (hbk)
ISBN 0-415-05621-7 (pbk)

To the memory of Rupert Shephard,
my father and friend

Contents

Tables

APPENDIX

Acknowledgements

This book is the voyage of discovery I have made over the past decade, during which time gender planning has dominated my life. Without the intellectual stimulation and emotional support of family, friends, students and colleagues in many different parts of the world, I would never have reached this stage. It is not possible to acknowledge by name all those who have influenced or helped me. I would, however, like to mention a few upon whom, knowingly or unknowingly, so much of this book has depended.

Norman Long, Alison MacEwen Scott and Bryan Roberts retained a stalwart belief in my academic credibility as I deviated from the mainstream; Marianne Schmink first told me of the triple role in a car journey from Manchester to London; Maxine Molyneux's articulation of gender interests I first heard at an IDS seminar; Caren Levy as my colleague and co-trainer from 1984 to 1990 collaborated and supported me through so much of the early development of gender planning; Diane Elson and Francis Stewart sharpened my understanding of structural adjustment; students, first at University College London's Development Planning Unit, and then at the London School of Economics and Political Science, taught me about gender and development from their perspective; Helen Doyle, Tovi Fenster, Lulu Gwagwa, Anna Robinson and Mary Jennings challenged many of my ideas; Mary Curran and Alicia Herbert provided research support.

Brian Moser first filled me with his enthusiasm for South America and introduced me to Guayaquil; Michael Cohen insisted I return to Guayaquil, and later provided an opportunity to start introducing gender planning into the Urban Division at the World Bank; Rosa Vera, and my 'comadres' Lucy Savala and Carmelina Guiterez taught me their survival strategies; above all, Emma Torres for more than a decade has shared so generously her home and her knowledge and love of Indio Guayas and has tried so patiently to ensure that I was 'enseñado'.

Margaret Legum and Sophia Tickel first encouraged me to develop NGO gender planning training courses; Fiona Thomas and Tina Wallace shared their early experiences; Judith Bruce and Aruno Rao worked to develop the field of gender training; Kate McKee supported my ideas in the early, hostile climate; Amelia Fort was brave enough to pioneer gender planning training with South American government ministries; Carole Hannan-Andersson and Brita Ostberg encouraged me to be innovative with the planning methodology.

Liz Barat, Monica Hicks, Marnie Piggot, David Satterthwaite and Nadia Taher at various points propped me up in London; Lourdes Beneria, Lewis Cohen, Monique Cohen, Pam Sparr and Sally Yudelman did the same in America; Ben Shephard kept in communication; Heather Gibson, my commissioning editor, never doubted and patiently waited.

Writing this book at times has been a painful experience. Various people, in particular, helped me when I was ready to give up. Michael Safier challenged me to develop my ideas on gender planning; Sukey Field through her belief in my work provided sustaining support; Rosalind Eyben with her qualities as an academic practitioner commented incisively on draft chapters; without Linda Peake's painstaking critique and editing work, this book would not have been completed.

Finally, my sons Titus and Nathaniel spent their teenage years soaking up gender planning, never complained when I 'closed the kitchen', and were always proud of me; and lastly, my partner throughout this voyage, Peter Sollis, never lost faith, was relentless in his encouragement and bullied me to finish. To all these and many more besides, I owe a great debt of gratitude.

Washington, DC

1 Introduction

This book describes the development of gender planning as a legitimate planning tradition in its own right. The goal of gender planning is the emancipation of women from their subordination, and their achievement of equality, equity and empowerment. This will vary widely in different contexts depending on the extent to which women as a category are subordinated in status to men as a category. The knowledge base explored in recent feminist and development debates provides the conceptual rationale for several key principles. These in turn translate into tools and techniques for a gender planning process. These analytical principles relate to gender roles and gender needs, also to control over resources and decision-making in the household, civil society and state. The procedures by which gender planning is operationalized as well as the framework within which it is institutionalized also require identification and acknowledgement before this new planning tradition is to attain legitimacy. For such a new planning tradition there is still a long way to go – what follows can only be a starting point documenting the development of gender planning over the last decade.

Why should the issue of gender constitute a legitimate planning tradition in its own right? To answer such a question it is necessary to start by examining both the current agenda of 'women in development' and the planning preoccupations of those involved in developing countries. For in the world of policy and planning where fashions come and go, women and development concerns are a peculiar anomaly. They resolutely refuse to disappear. However, unlike other recent contenders, such as the environment, they have not succeeded in attaining planning legitimacy. Why has it been so easy for environmental planning to gain identity as a separate planning tradition, and yet so difficult for the 'women in development' approach? Why do the proliferating numbers of policies and plans of action for women still only too frequently fail to be translated into practice? Why are Women's Ministries so effectively excluded from national planning processes and marginalized in terms of resource allocation?

The background to these questions is well known. The United Nations' Decade for Women (1976–85) played a crucial part in highlighting the important but often previously invisible role of women in the social and economic development of Third World countries and communities, and the particular 'plight' of low-income women. During this decade there were considerable shifts in approaches both by academic researchers and by policy-makers. Researchers moved away from a preoccupation with the role of women within the family and women's reproductive responsibilities, towards an understanding of the complexities of women's employment and their productive activities. Research on both waged workers and those in the informal sector, in urban and rural areas, helped in identifying the range of low-income women's income-generating activities in Third World economies. Equally, during the decade policy-makers began to shift their focus from a universal concern with welfare-orientated, family-centred programmes which assumed motherhood as the most important role for women in the development process, to a diversity of approaches emphasizing the productive role of women. Despite developments such as these highlighting the importance of woman to the development process, the acceptance of gender planning has been hindered by a range of issues which still require clarification.

CONCEPTUAL ISSUES: FROM SEX OR GENDER TO WID OR GAD

The term 'women in development' was coined in the early 1970s by the Women's Committee of the Washington, DC, Chapter of the Society for International Development, a network of female development professionals who were influenced by the work on Third World development undertaken by Ester Boserup and other 'new' anthropologists (see Boserup 1970; Tinker 1982; and Maguire 1984). The term was very rapidly adopted by the United States Agency for International Development (USAID) in their so-called Women in Development (WID) approach, the underlying rationale of which was that women are an untapped resource who can provide an economic contribution to development. USAID, with its Office of Women in Development, has been one of the most determined advocates of the WID approach. Together, with the Harvard Institute of International Development, they have produced a case-study based methodology to identify how women have been left out of development on the grounds that 'women are key actors in the economic system, yet their neglect in development plans has left untapped a potentially large contribution' (Overholt *et al.* 1984: 3).

More recently a further shift in approach, principally in academic research, has recognized the limitations of focusing on women in isolation and has drawn attention to the need instead to look at 'Gender and Development' (GAD). This focus on 'gender' rather than 'women' was influenced by such writers as Oakley (1972) and Rubin (1975). They were concerned about the manner in which the problems of women were perceived in terms of their *sex* – namely, their biological differences from men – rather than in terms of their *gender* – that is, the social relationship between men and women, in which women have been systematically subordinated.[1]

Approaches to issues relating to women in developing countries became concerned therefore with the manner in which gender and concomitant relationships were socially constructed. The focus on gender rather than women makes it critical to look not only at the category 'women' – since that is only half the story – but at women in relation to men, and the way in which relations between these categories are socially constructed. Men and women play different roles in society, with their gender differences shaped by ideological, historical, religious, ethnic, economic and cultural determinants (Whitehead 1979). These roles show similarities and differences between other social categories such as class, 'race', ethnicity and so on. Since the way they are socially constructed is always temporally and spatially specific, gender divisions cannot be read off on checklists. Social categories, therefore, differentiate the experience of inequality and subordination within societies.

Although the critical distinction between sex and gender is well known, the further distinction between Women in Development (WID) and Gender and Development (GAD) is less clear. The terms are all too often used synonymously, yet in their original meaning they are representative of very different theoretical positions with regard to the problems experienced by low-income women in the Third World. Consequently, they differ fundamentally in terms of their focus, with important implications for both their policies and planning procedures.

The WID approach, despite its change in focus from one of equity to one of efficiency, is based on the underlying rationale that development processes would proceed much better if women were fully incorporated into them (instead of being left to use their time 'unproductively'). It focuses mainly on women in isolation, promoting measures such as access to credit and employment as the means by which women can be better integrated into the development process. In contrast, the GAD approach maintains that to focus on women in isolation is to ignore the real problem, which remains their subordinate status to men. In insisting that women cannot be viewed in isolation, it emphasizes a focus on gender relations, when designing measures to 'help' women in the development process.

At the beginning of this book it is important to recognize that gender planning differs fundamentally from planning for Women in Development.

Because it is a less 'threatening' approach, planning for Women in Development is far more popular. However, by its very definition it is an add-on, rather than an integrative, approach to the issue. Gender planning, with its fundamental goal of emancipation, is by definition a more 'confrontational' approach. Based on the premise that the major issue is one of subordination and inequality, its purpose is that women through empowerment achieve equality and equity with men in society.

PROCEDURAL ISSUES: FROM SOCIAL AWARENESS TO PLANNING PRACTICE

If the first problem is conceptual categories, the second is planning procedures. It is clear that while the important role that women play in Third World development processes is now widely recognized, conceptual awareness of both WID and GAD has not necessarily resulted in their translation into planning practice.

The extent to which social awareness or consciousness of inequality has been satisfactorily incorporated into planning varies widely. Class inequality, for example, has probably most commonly been addressed in planning through the introduction of income as a target group indicator for selection in policies, programmes and projects. The fact that the translation of social categories into planning indicators is neither automatic nor universal has had important implications for those addressing the concerns of gender inequality. Indeed, for many practitioners involved in different aspects of development planning, the lack of an adequate gender planning methodology has been the most problematical aspect of their work.

With the endorsement of the 1985 UN Forward Looking Strategies, institutions at international, national and non-governmental level now play lip-service to WID. Mostly focusing on women in isolation, Ministries of Women's Affairs, WID Units, and Women's Non-Governmental Organizations (NGOs), with predominantly female staff, have proliferated throughout the world, in countries as diverse as the Philippines, Zimbabwe and Belize. They have been involved in developing WID policies, in designing WID checklists and in formulating programmes and projects to 'bring' or 'integrate' women into the planning process. More recently, some organizations have moved towards a more 'gendered' approach, with many embarking on extensive training programmes aimed at changing the work practices of colleagues within the organization.

Despite the energy and resources allocated to this work for more than a decade, WID still most frequently remains an 'add-on' to mainstream policy and planning practice. It continues to experience difficulty in being satisfactorily incorporated into the diversity of sectoral interventions concerned with the lives of low-income communities in Third World countries. There are certainly important success stories. Nevertheless, in a context where so many policies fail to become translated into practice, the fundamental preoccupation remains the need to develop a far more rigorous planning framework than that which currently exists. Only then will gender concerns be integrated into development practice.

A number of problems have contributed to the failure to develop a gender planning framework. First, most authorities responsible for development planning have only reluctantly, if at all, recognized gender as an important planning issue. Despite the creation of women's ministries, units and bureaux, decision-making powers largely remain male-dominated and gender-blind. The political constraints are overwhelming. However, to focus entirely on these is to miss a number of other crucial problems, some of them more technical in origin.

To begin with, the majority of policy-makers and practitioners working on WID/GAD issues do not themselves have any formal training in the discipline of planning. It is interesting to note that for planners working within such traditions as transport, land-use or regional planning, for example, comprehensive training is an assumed prerequisite; yet this is not the case with gender planning. Here the tendency is to recruit women on the basis that they will inherently understand the issues and rely on their good 'common sense'. In many cases this has resulted in widespread ignorance of the in-built limitations of planning procedures adopted.

A further problem is that the concern of feminist academic research, by its very nature, has been to highlight the complexities of gender relations and divisions of labour in specific socio-economic contexts. It has not been concerned to identify how such complexities might be simplified into methodological tools which enable practitioners to translate gender awareness into practice. The audience of pure research still remains essentially other academics. This failure to translate the results of research into practice means that many of those committed to integrating gender into their work at policy, programme or project levels still lack the necessary planning principles and methodological tools. This issue is critical; planners require simplified tools which allow them to feed the particular complexities of specific contexts into the planning process.

Finally, and of greatest importance for those who are involved in planning practice, it has proved remarkably difficult to 'graft' gender onto existing planning disciplines. These have proved remarkably resistant to change. It is probably no coincidence that gender planning, as a planning discipline in its own right, developed out of a planning institution rather than an academic, research or consultancy environment. The personal frustration I experienced when teaching on a variety of training courses for Third World and bilateral planning practitioners, which I tried to make gender-aware by 'grafting' gender onto particular planning disciplines – be it land-use, 'manpower' or infrastructure planning – led me to recognize the necessity of distinguishing between a gender-aware planner (in, say, transport planning), and gender planning, as a specific planning approach in its own right.

As a consequence of factors such as these, women and gender remain marginalized in planning theory and practice, and will do so until such time as theoretical feminist concerns are adequately incorporated into a policy and planning framework, which is recognized as a planning tradition, with its own planning methodology.

DEFINITIONAL ISSUES: POLICY, PLANNING AND THE ORGANIZATION OF IMPLEMENTATION

It is important to realize that the problems of integrating WID or GAD vary at different stages in the planning process. This makes it necessary to define how such terms as 'policy', 'planning' and 'implementation' are used throughout the book. This is of particular concern given the widespread confusion in existing terminology, and the fact that the close interrelationship between them makes it difficult to distinguish where one term ends and another begins (Conyers 1982). If *policy* is about *what to do*, then *planning* is about *how to do it*, the *organization of implementation* is about *what is actually done*. The term 'planning process' is used generically to describe the three stages, outlined below, in what essentially is a continuous process.

- Policy-making: the process of social and political decision-making about how to allocate resources for the needs and interests of society, concluding in the formulation of a *policy strategy*.
- Planning: the process of implementation of the policy, often concluding in a *plan*.
- The organization of implementation: the process of administrative action to deliver the programme designed, often resulting in a completed *product*.

Similarly, the term 'gender planning process' is also used generically to describe the three interrelated stages of gender policy, gender planning and the organization of implementation, with the term 'gender planning methodology' referring to the detailed methods by which the process is achieved.[2]

The distinction between different stages in the planning process is critical. For instance, where there is gender-blindness in policy formulation one of two problems is likely to occur. First, women are not recognized as important in development processes and simply not included at the level of policy

formulation. Secondly, development policy, even when aware of the important role women play in development processes, because of certain
assumptions, often still 'misses' women, and consequently fails to develop
coherently formulated gender policy.

In contrast, the inability to translate gender policy into implemented
practice is often a different problem. This relates to identifying constraints
that can occur in numerous phases in the implementation stage. In planning,
the term 'culture' is frequently used as a blanket explanation to identify
constraints in planning procedures. The use of this term in such a pejorative
manner, as a causal explanation of failure, raises the issue of the extent to
which planning is a neutral activity. Are the planning constraints encountered
when challenging inequalities in society more often political rather than
technical in nature?

POLITICAL AND TECHNICAL ISSUES: CONSTRAINTS AND
OPPORTUNITIES IN THE PLANNING PROCESS

Is gender policy not implemented because of such technical constraints as
inappropriate planning procedures, or are there wider political constraints,
operating at the level of policy formulation, which impede successful implementation? Grindle (1980) identifies the manner in which social and
structural constraints influence the 'implementability' of programmes, ar-
guing that people and groups aiming to transform social relationships
generally meet with opposition from groups whose interests they threaten.
She maintains that the content of policy has considerable impact on the kind
of political activity stimulated by the policy-making process. She claims that
implementation is an ongoing process of decision-making by a variety of
actors, the ultimate outcome of which is determined by the content of the
programme being pursued and by the interaction between decision-makers
within a given politico-administrative context.

In the same way the traditional view of planning methodology as a neutral
and universally applicable set of technical procedures has been criticized.
The characterization of planning as contentless and contextless, with an
'appearance' of neutrality, according to Thomas (1979), was the result of the
separation of the conception of planning from the concrete reality of its
context. The methodology of planning as a trans-historical, apolitical and
technical set of procedures has been criticized as 'empirically vacuous' (Scott
and Roweiss 1977), as has been the assumption that since planning is
primarily identified as a public-sector activity, its institutions are 'neutral',
and acting in the 'public good' (Healey *et al.* 1982). In examining the
implementation of gender policy it is, therefore, important to identify the
extent to which constraints are technical or political in nature. This distinction

can assist in a more realistic assessment of the 'room for manoeuvre' within specific contexts, and in more accurate explanations as to why changes to WID/GAD policy so frequently occur at specific stages of the planning process.

GENDER PLANNING AS A NEW PLANNING TRADITION: THE STRUCTURE OF THE BOOK

The book is divided into two parts, reflecting stages in the gender planning process. Part One, The Conceptual Rationale for Gender Planning in the Third World, examines feminist theories and WID/GAD debates in terms of their relevance for gender planning. Its concern is with the gender-blindness of current policy formulation and planning procedures. It identifies fundamental misconceptions and assumptions which have caused development planners, even if inadvertently, to discriminate against, or miss out, women.

Part One consists of four chapters. Chapter 2 examines assumptions relating to family structure, to divisions of labour within the household and to the household as a natural joint decision-making unit. These provide the conceptual rationale for the key principles of gender planning methodology. This concerns the fact that men and women play different roles in society, have different levels of control over resources, and therefore often have different needs.

Chapter 3 describes the concept of gender interests, and their translation into planning terms as gender needs. An important distinction is made between practical and strategic gender needs. These needs are discussed in relation to the way in which the state, in different political contexts, effectively controls women's strategic gender needs through family policy relating to domestic violence, reproductive rights, legal status and welfare policy. The usefulness of these gender planning tools is then examined in terms of different interventions in planning sectors such as employment, human settlements and housing, the environment, and basic needs such as transportation and water.

Chapter 4 discusses the interrelationship between different macro-economic development models and policy approaches to Third World women. It shows the extent to which particular emphases on gender have determined different women and development policies, identifying a shift in policy approach from 'welfare' to 'equity', then from 'anti-poverty' to 'efficiency', and finally to 'empowerment'. These shifts have taken place, not in isolation, but have mirrored general trends in Third World development policy, moving from modernization policies of accelerated growth, through basic needs strategies associated with redistribution, to the more recent structural adjustment policies.

This analysis provides the conceptual framework for the gender planning tradition and methodology. This is described in Part Two of the book, the Gender Planning Process and the Implementation of Planning Practice. Chapter 5 characterizes the emerging planning tradition of gender planning, and outlines its methodological tools, procedures and components. The knowledge base for this new planning tradition comes from sources such as the feminist debates and the recent WID/GAD debates. Its goal is the emancipation of women and their release from subordination, and its specific aim identified as the achievement of equality, equity and empowerment through the meeting of strategic gender needs. Chapter 5 also identifies the integration of such tools as gender-needs assessment and the WID/GAD policy matrix into such procedures in the gender planning process as gender diagnosis, gender objectives and entry strategies.

Frequently, the most important problem faced by planning practitioners is their inability to translate gender policy into practice. In addition, in the complex reality of planning processes it is very difficult to separate out the many different components which determine the implementation process. These can be institutional structures, organizational procedures or planners' attitudes. However, there is a tendency to see constraints and opportunities deterministically as either technical or political in nature. The last four chapters of the book focus on different components in the planning process, while recognizing that in the reality of planning practice they are highly interrelated.

When a new issue appears on the political agenda, the first stage in its resolution is often the creation of a new institutional structure. In the case of gender, for instance, the formulation of a policy does not necessarily mean that this is institutionalized into existing planning agencies. Chapter 6, therefore, focuses on institutional components. One critical debate concerns the extent to which it is more satisfactory to institutionalize gender within existing mainstream planning organizations, or whether it requires specially formed structures, such as ministries of women's affairs in national governments, women's departments in donor agencies and women's units in NGOs.

The formulation of a gender policy does not mean that it gets operationalized, since severe problems are often experienced by those trying to translate it into practice. Chapter 7 focuses on operational procedures in the planning process. These are identified as the 'technical' domain of planning, in which the major problem identified has been a lack of adequate procedures. Here the debate concerns the extent to which constraints are technical in nature, relating to inappropriate planning procedures, or whether there are wider political constraints which impede successful implementation. The chapter reviews the range of new planning procedures to operationalize gender concerns introduced during the past decade by organizations at international, national and NGO level. The analysis highlights constraints in such procedures, with a detailed examination of the introduction of gender into the project planning cycle.

Training is another critical component to ensure successful integration of gender planning into practice. Chapter 8, therefore, provides a description and critique of the role that it plays in creating gender awareness and sensitivity, providing practitioners with appropriate tools and techniques. It outlines different training methodologies and identifies the different components of a gender training strategy. Finally, it highlights the successes and limitations experienced by a diversity of institutions which currently utilize training. In conjunction with this chapter, the Appendix outlines the particular training strategy and methodology associated with gender planning. In describing its objectives, the workshop programming and format, contents structure and materials required, the intention is to provide practitioners with the necessary material for implementing training sessions.

Because gender planning is not an end in itself, but a means by which women, through a process of empowerment, can emancipate themselves, Chapter 9 concludes by placing gender planning in its wider political context. Since the development of gender planning comes out of the social and political movements that women themselves now generate, rather than because of state intervention, its success ultimately must depend on their participation in the planning process. This final chapter, therefore, examines current Third World women's organizations and movements. It describes the mixed experience of such organizations in raising consciousness to confront women's subordination, creating alliances and linkages to ensure the success of planning processes. Finally, it highlights entry points identified by women's NGOs for negotiation and debate around women's needs at four different levels: household; civil society; the state; and the global system.

THE IMPORTANCE OF GENDER PLANNING IN THE CURRENT CONTEXT

The development of gender planning as a planning tradition is of critical importance for a number of reasons. Probably of greatest importance is the urgent need to inform policy, through the formulation of gender policy at international, national and NGO levels, as well as its integration with sectoral planning. In addition, it is needed to ensure the development of more appropriate – that is, gender-aware – planning procedures. Finally, it can assist in the clarification of both technical and political constraints in the implementation of planning practice.

The particular importance of developing gender planning as a rigorous new planning discipline in its own right lies in the fact that the 'WID business' is now in crisis. Although the resources allocated to WID or GAD programmes and projects in reality have never amounted to much, the propaganda surrounding them meant that to many this was seen as a much-resourced sector. The pressure from the 1976–85 Decade, which resulted in the creation of institutions with a particular focus on women, and the allocation of resources to their needs, was replaced at the end of the 1980s by a more hostile climate of debt and recession in which the overriding preoccupation of national governments, donor agencies and NGOs alike is with efficiency in the allocation of resources. At the same time it is clear that severe problems have been experienced in operationalizing gender issues – namely, ensuring that formulated policies become implemented practice. These two factors, the largely unsuccessful track record and the changing global economic context, make it essential for those involved in WID or GAD issues, whether as academics or practitioners, to understand more comprehensively gender policy and planning processes and consequently the constraints on the implementation of practice. The purpose of this book, therefore, is to assist in this process.

Part I

Conceptual rationale for gender planning in the Third World

2 Gender roles, the family and the household

Can we plan for the needs of low-income households generally or is it necessary to plan for the needs of women in their own right? This very simple question provides the basis for the development of gender planning, as a new planning discipline with a specific focus on the issue of gender. It also allows us to recognize that because women and men have different positions within the household and different control over resources, they not only play different and changing *roles* in society, but also often have different *needs*. It is this role and needs differentiation that provides the underlying conceptual rationale for gender planning and defines its long-term goal as the emancipation of women. The fundamental planning principles, therefore, derive from the need to disaggregate households and families on the basis of gender, when identifying planning needs.

The development of gender planning as a planning tradition in its own right is the consequence of the inappropriateness of current planning stereotypes. These relate to the structure of low-income households, to divisions of labour within them, and to power and control over resource allocations between different members in such households. Despite the common rhetoric of 'planning for people', in much of current policy and practice, based on Western planning theory, there is an almost universal tendency to make three broad, generalized assumptions, despite the empirical reality of the particular planning context:

1 that the household consists of a nuclear family of husband, wife and two or three children.
2 that the household functions as a socio-economic unit within which there is equal control over resources and power of decision-making between all adult members in matters influencing the household's livelihood.
3 that within the household there is a clear division of labour based on gender. The man of the family, as the 'breadwinner', is primarily involved in productive work outside the home, while the woman as the housewife

and 'homemaker' takes overall responsibility for the reproductive and domestic work involved in the organization of the household.

In many societies these Western planning stereotypes of household structure and the gender division of labour within it are seen to reflect the 'natural' order. Consequently, the state and family ideologically reinforce them through the legal and educational system, the media and family-planning programmes, without recognition that within the family the woman's position is subordinate to that of the man. Moreover, such an abstract stereotype model of society has severe limitations when applied to most Third World contexts.

In this chapter examination of each of these three assumptions provides the basis for developing a gendered approach to planning. In reality, these three assumptions interrelate. Their division, while essential for the development of a gender planning methodology, is mechanistic. Their separation reflects differences in the priorities of planning disciplines as well as changing preoccupations with gender issues within planning itself.

THE HETEROGENEITY OF HOUSEHOLD STRUCTURES

The first assumption is that the predominant household structure consists of a nuclear family of husband, wife and two or three children. In reality this is no more than an idealized traditional planning stereotype even of Western industrial society today. Yet the failure to recognize that low-income households are not homogeneous in terms of family structure is still widespread. Although nuclear families may be the dominant type in some contexts, other structures exist. This is the case not only in industrialized countries but also in the developing world. Planners trained in European and North American planning schools have successfully exported such assumptions (Pascall 1986).[1]

Linked to the model of the nuclear family as the basic unit of society is the concept of headship. The idea that a 'head', normally assumed to be a man, represents and manages the household, Rogers argues, is a 'figment of the statistician's imagination' (1980: 64). In his definition as 'head of household', the man as the breadwinner is perceived to be the financial supporter, with all other members defined as 'dependants'. This is the case in both rural and urban contexts, when the woman is the primary income-earner, and it is the man who more accurately should be defined as a dependant. Frequently women are only counted as heads when it becomes impossible to list the oldest male present, be it the grandfather or grandson, because there is no likely candidate. In her detailed description of the treatment of women in quantitative statistics, Rogers (1980) provides chilling, if salutary, examples of the manner in which the data-collection categories in different agricultural projects, by making women's agricultural participation invisible, effectively excluded them from land, credit, extension and other services. The designation of headship in national household surveys has been questioned. However, it continues to be widely practiced with the assumption that the attitudes and motivations of all members of the household are identical with those expressed by the male household head.[2]

More recently, several non-nuclear family structures have reached the attention of policy-makers. It is now accepted that the extended family does not necessarily disappear with 'modernization' or 'urbanization'. Where it remains vital for low-income survival strategies, as happens in countries experiencing stringent adjustment conditions, extended families of widely different and complex structures may not only survive but increase in numbers.[3] However, the most important non-nuclear family household structure still 'invisible' in many planning contexts is the female-headed household. This was first noted by researchers in the Caribbean, but now is recognized as a global phenomenon. Among the very wide range of households, two main types have been identified. First, *de jure* female-headed households, in which the male partner is permanently absent due to separation or death, and the women is legally single, divorced or widowed. This is common in situations of urban poverty. Second, *de facto* women-headed households in which the male partner is 'temporarily' absent. Here the woman is not the legal household head. She is often perceived as a dependant although she may, for most of her adult life, have primary if not total responsibility for the financial as well as the organizational aspects of the household.

Two very different socio-economic and political conditions have resulted in the increase in the numbers of *de facto* female-headed households during recent decades. Most important of all is male labour migration. This consists not only of the rural–urban migration common in parts of Africa, but also the dramatic flows of international migration such as those from the resource-poor areas of Asia to the rich Middle Eastern economies. Female-headed households also emerge under conditions of war, insecurity and disaster, whether 'man-made' or natural.

It is estimated that women head one-third of the world's households. In urban areas, especially in Latin America and parts of Africa, the figure reaches 50 percent or more. In rural areas where men traditionally migrate it has always been high, while in refugee camps in areas of Africa and Central America it is nearer 80 to 90 percent. While there are considerable regional variations, the number of *de facto* female-headed households is increasing rather than declining. In many parts of the world this is not a new phenomenon, simply one that is now more openly acknowledged. The reasons for this recognition are twofold. First, attitudes within communities are changing about the stigma of a woman living without a man. This is particularly evident in communities under conflict, where women forced to live on their own have themselves become more confident in confronting criticisms about their status (Weeda 1987). Secondly, there is growing recognition by planners implementing projects, if not formulating policy, that they cannot ignore women who head households if they are concerned with the successful completion of their work.[4]

The economic conditions of female-headed households vary considerably. They depend on such factors as the woman's marital status, the social context of female leadership, her access to productive resources and income, and the composition of the household. Some *de facto* female-headed households in which husbands regularly send remittances home are better off than if their husbands were present. In other cases, where men become 'guest husbands' with a series of temporary liaisons, the situation is worse. Nevertheless, it is important not to assume *a priori* that households headed by women are poorer than those with a male 'head'. Clearly, women choose not to marry in some contexts, and in others to marry but live separately from their husband.

The crucial issue here relates to the apparent trend of increasing detachment of women and children from men's income. This results not only from changes in legal arrangements or sexual practices, but because of the economic cost of bearing children. Where these households have a higher dependency ratio and a lower participation rate among potential earners, as well as earners with less income-earning power, all too often they fall below the poverty-line, and are over-represented at the lower end of the income distribution, disproportionately numbered among the poorest of the poor (White *et al.* 1986). Female-headed households in poverty are poor for many of the same reasons as male-headed households. It is important, therefore, to identify whether it is useful to target female-headed households as a separate category. In reality, the most disadvantaged are those households without adult income-earners other than the mother, and with many dependent children. In such cases children are less likely to attend school and more likely to work with obvious consequences for the inter-generational transfer of poverty. Women who head households are not a homogeneous category. Nevertheless, the problem of balancing multiple roles is exacerbated for this group, which can have policy implications in specific contexts.

THE HOUSEHOLD AS A NATURAL DECISION-MAKING UNIT

The second assumption, and one most commonly made by macro-economic planners working on problems of rural development, is that the household functions as a natural socio-economic unit, within which there is equal control over resources and power of decision-making between adult household members in matters influencing their livelihood. This is based on two critical premises that require detailed examination from a gender perspective.

Families and households

The first premise, at the inter-household level, is that of the household as a natural socio-economic unit. This simple but fundamental planning stereotype derives from the assumption that the family, a social unit based on kinship, marriage and parenthood, and the household, a residential unit based on co-residence for such purposes as production, reproduction, consumption and socialization, are 'naturally' and universally synonymous units. In this 'one pot, one roof model' (Lehmann 1986) the family live and work together as the basic labour unit, sharing both work and the proceeds of their labour. As Young (1990) has argued, this is based on a conflation of the two separate concepts of households and families. Many assumptions made about the nature of intra-household relations relate to marriage relationships and relations between parents and children.

The idea that the household functions as a single socio-economic unit, organized as an independent entity with clearly defined boundaries that separate it from other households in the socio-economic structure in which it is located, is not borne out in reality (Evans 1989). Although families and households may overlap in some societies, particularly those that are Western and urban, in others they do not. Wide variations in marriage and kinship systems influence residential and conjugal arrangements. These vary both spatially and temporally. Internal factors relating to the nature of the marital contract, patterns of inheritance and the different stages of expansion, consolidation and contraction in the family life cycle influence them. In addition a diversity of external socio-economic and political factors are also important (Whitehead 1984a, 1984b; Guyer and Peters 1987). Whereas planners treat households as static units, in reality they vary widely depending on socioeconomic contexts. These are themselves subject to constant restructuring, with their limits and boundaries often difficult to find out.

The concept of the household as a unified economic entity obviously fails to recognize inter-household resource and labour exchanges and systems of reciprocity. The most formalized are those that exist in polygamous households, but they are also widespread in other family structures. Of particular importance for women are the various forms of co-operation and collectivity in reproductive work *between* households. A common form is where some women undertake such domestic work as the minding of children and cooking or fuel and water collection, in order to release others to engage in wage labour or subsistence production (Harris 1981). Such arrangements may be institutionalized through kinship ties, or else based on informal arrangements of solidarity to meet a joint need.

The household as a joint utility function[5]

A further tendency of household-level analysis, that of looking *at* households rather than *inside* them, ignores the importance of intra-household dynamics. The second premise, at the intra-household level, that requires particular scrutiny from a gender perspective, concerns the household as a joint decision-maker. This relates particularly to questions of power and control over the distribution of resources between household members in matters affecting their livelihood.

The household has long been used as the unit of analysis for census, statistical and survey purposes. Since the mid-1960s, however, it has gained a particular prominence among macro-economic planners, because of their increasing concern to develop and apply a model of household decision-making behaviour for several diagnostic and predictive purposes. A number of models of the household as unified undifferentiated units have been developed from very different political perspectives, influenced above all by the work of Chayanov and Becker. From a Marxist perspective, for instance, Chayanov's Theory of Peasant Economy has been used to develop models of household utility maximization under conditions of land and labour constraints. In this the tightly bounded household unit is identified as the operative unit of production and consumption (Lehmann 1982). In contrast, Becker (1965) has combined arguments about the economic rationality of household behaviour with the neo-classical theory of the firm to develop a new theoretical approach, known as New Household Economics (NHE). Here a complex array of relationships and exchanges within households are collapsed into a function that is similar to that of an individual decision-maker.

As Evans (1989) has argued, models based on NHE identify the household, rather than the individual, as the most relevant unit of 'utility maximization'. The family is identified as the basic unit not only of production but also of consumption. Its utility derives not simply from the consumption of goods and services purchased in the market place, but also from the range of home-produced goods and services, the so-called 'Z-goods'. Because they comprise objects of final consumption produced within the household, such Z-goods are often considered similar to use-values. They are goods produced within the household with fixed capital, variable inputs and labour, combined with different levels of labour and capital intensity. The particular combination will depend not only on the level of household technology, but also on the prices of market goods, an imputed rate of return to household capital, and the shadow price of time – in other words, the domestic worker's earnings forgone from the labour market (Hart 1990). Thus the household does not simply maximize profits but rather it maximizes the joint utility of all its members. In addition, welfare maximization is based on the condition that the 'free choices' or preferences of household members are equally represented in the household utility function, such that the welfare of each family member is normally integrated into a unified family welfare function (Schultz 1988). These assumptions will now be examined in more detail. Among the many limitations from a gender perspective of the notion of the household as a joint utility function, Evans (1989) has identified three of particular importance.

Households as unified units of consumption and production

The basic premise of the NHE model that households are unified units of consumption and production with all choices made by the same decision-making family unit is not borne out by empirical evidence. In reality, in the same way that families and households do not necessarily overlap, so neither are production and consumption units necessarily unified at the household level. The unit of production in both agricultural cultivation, as well as in informal sector activities, often extends beyond the household to include others. These are linked by such criteria as kinship, ethnicity, gender or geographical location. Similarly, units of consumption often expand to include extended family or neighbours. In such cases consumption decisions are not necessarily made jointly after all production decisions are made. They are more likely to be based on the different production possibilities of various household members.

The joint utility function and the substitutability of labour

The second important limitation of the NHE is its explanation of the household division of labour in terms of comparative advantage. The NHE argument is that since individual labour time is valued in terms of a market wage, it is obviously allocated to tasks in which it is more efficient. Labour is assumed to be perfectly flexible and substitutable, with the gender division of labour between women's household work and men's wage work legitimized as both functional and efficient for household welfare maximization. Low (1986), for instance, argues that it is to the mutual advantage of the 'rational' farm-household to allocate women's labour time to household-based activities while encouraging men to migrate to off-farm employment. However, as Evans has commented, 'This explanation completely obscures the non-economic and ideological factors that discriminate between male and female labour in the market-place, and values their labour differently in market and non-market sectors' (1989: 10). In reality, family labour time is not like other factors of production that can be flexibly allocated based on its comparative costs in market and non-market activities. Gender as well as age and status are all critical determinants in differentiating the mobilization and allocation of family labour to different activities. Not only do the divisions of labour based on gender define reproductive work as women's work but they also segregate the productive work undertaken by men and women in both agricultural and industrial sectors.

The assumption in NHE concerning the substitutability of labour has had critical implications during the past decade in relation to policy measures accompanying structural adjustment loans. The incentives introduced by governments to encourage farmers and entrepreneurs alike to switch from non-tradable to tradable goods have been based on the assumption that households would respond to new opportunities by reallocating their labour (Elson 1991). In rural areas increased production of export-orientated crops has resulted in increased productive work for women in their roles as peasants. This has been accompanied by a reduction in their subsistence production for household consumption (Feldman 1989). In urban areas it has meant greater unemployment for men displaced from jobs producing non-tradable goods. For women drawn into export-orientated manufacturing or undertaking increased domestic service to supplement family income this has meant extra paid work added to their unpaid reproductive work (Moser 1992a). Where women's preferences in terms of allocating labour do not necessarily concur with that of the household head, conflict is a likely outcome (Whitehead 1984b).

The joint utility function, decision-making and resource allocations

The third limitation of the NHE concerns the use of the joint utility function to deal with the issue of preference aggregation in relation to both decision-making and resource allocation. Critical here is the fact that welfare maximization is conditional on the 'free choices' or 'preferences' of household members being equally represented in a single household utility function. How do the single family welfare functions cope with the fact that within households individuals are perceived to make free and voluntary economic choices? This has proved problematic for economists, particularly in relation to who makes the decision.

Becker (1965), for instance, originally used the concept of *altruism* to avoid confronting the issue of preference aggregation. He assumed that because individual choices are motivated by the desire to maximize total family welfare, household members subordinate individual tastes and preferences in pursuit of common goals. However, the identification of self-interest as the preserve of the market place, with altruism limited to the sphere of the family, was widely criticized as avoiding the issue of conflict within the family. To explain why individual family members do not 'free-load' on the benevolence of others, and how families formulate goals in the interests of all members, Becker therefore substituted the assumption of pure altruism with an assumption that a 'benevolent dictator' rules the household. In practice this is assumed to be the male head of family. He sets the goals of the household in the interests of the household as a whole, and ensures that conflict or inequality is eliminated by working in the interests of all parties. In this way the socio-economic interests of family dependants such as women and children are internalized within the utility function of the husband and father.

However, the male household head may not have any real understanding of the day-to-day problems associated with household welfare, since provisioning is a reproductive task of women. Men frequently know little of their wives' coping strategies. When men are absent from home in West Africa they remove themselves 'from the children's cries for food' (Whitehead 1981). In some cases, as Harris (1981) has identified, the authority of the male head does not necessarily coincide with the residential family unit, but may lie with kin elsewhere or with some wider economic or political sphere. It cannot be assumed that one altruistic member of the family, for whom the household head is a proxy, cares sufficiently about all members to transfer resources to them. Neither can it be assumed that the presence of an altruist induces other household members to act altruistically.

As Folbre (1986b) has identified, the fundamental problem with the concept of the joint utility function is that it removes the possibility of exploring conditions of unequal exchange and exploitation between family members. In particular, this is so between men and women, as they relate to decision-making and the allocation of resources. The assumption that the family pools and shares its resources derives not only from the conflation of families and households but also from the stereotypical assumptions planners have about marriage. This is identified as a 'sanctum protected from the conflicts that characterize most other social institutions' (Dwyer and Bruce 1988: 3). This conception of marriage as a partnership between two persons with reciprocal but unequal rights and obligations to each other, with similar rights over and obligations to the children of the marriage, and with a common set of interests, has resulted in the assumption that marriage produces a unit marked by joint control and management of resources (Young 1990).

Yet ideological and cultural as well as economic reasons underlie the symmetries and asymmetries in intra-household resource allocation. It is questionable whether family labour, income and resources, on the grounds of allocative efficiency, enter into a common fund. Equally controversial is the idea that they are distributed equitably such that all family members have access to pooled resources sufficient to meet their personal and collective needs. Empirical research clearly shows that unequal exchange and inequality exist within households. Intra-household decision-making, management and distribution arrangements vary depending on the household form and the nature of the 'conjugal contract' (Dwyer and Bruce 1988). The household is not necessarily a 'collectivity of mutually reciprocal interests' (Whitehead 1984a). Even where an ideology of sharing exists this does not necessarily mean that an equal distribution of resources occurs. Within the household, even in non-market environments, self-interest is often the predominant motivation. Since the 'benevolent dictator' does not represent household needs, his welfare cannot be taken as proxy for the welfare of all household members.

This has been most vividly illustrated in the documentation on the distribution of resources within households. Here the subordination of needs based on gender is often clearly visible. Women regularly eat last, and less (with girl children also discriminated against in the allocation of food) (Chen *et al.* 1981), less frequently get new clothes, or go without luxuries. These are all examples, Whitehead (1984a) has argued, of the 'maternal altruism' that is part of women's obligation to the family. Although women and men often own, and have access to income and resources, this is generally structured differently. Although, wide regional variations exist in women's access to and control over land in Africa, Asia and Latin America, nevertheless in most rural societies women's access is largely indirect. They acquire land by means of their relationship to individual males such as husbands, fathers or brothers by virtue of their gendered roles as wives or mothers. Men, in contrast, own land in their own right or by virtue of their lineage membership or other systems of inheritance.

Men and women not only have differing access to resources. Gender-based responsibilities also result in differences in the management and distribution of resources within the household. Gender divisions of income allocation vary widely. Cultural traditions determine which aspects of collective expenditure each must cover. Rarely are women's and men's income allocated to the same expenditure categories. In some societies husbands are responsible for housing and children's education. While income for food and clothing can vary, ultimately, almost universally women allocate income to day-to-day food, clothing needs and domestic goods. The fact that they take primary responsibility for household provisioning, means far greater allocation of women's income than men's to everyday subsistence and nutrition (Blumberg 1988). Studies of 'good mothering' show that a child's nutritional level correlates positively with the size of a mother's income, food inputs from subsistence farming and the quality of family-based child care (Kumar 1977).

Other models of the household

Constraints in the capacity of NHE models to handle the dynamics of intra-household inequalities have resulted in the promotion of other models of the bargaining household. These recognize that family decisions are more properly modelled as negotiations among primarily self-interested individuals in which members contend and exchange to gain their individual ends (Bernheim and Stark 1988). Manser and Brown (1980) and others have used the work of Nash (1953), because his model of both conflict and co-operation has been identified as particularly relevant to the analysis of household decisions. Household budgets are managed and distributed in a variety of ways, not always obviously rational in economic terms and with varying amounts of conflict and co-operation between family members. Sen (1990), in his interpretation of this model, defines households as experiencing 'co-operative conflict'. He argues that conflicts of interest between men and women are very unlike other conflicts, such as class conflicts. A worker and a capitalist do not typically live together under the same roof – sharing concerns and experiences and acting jointly. It is this aspect of 'togetherness' that gives the gender conflict some very special characteristics. The assumption of clear and unambiguous perceptions of individual interests, Sen maintains, misses crucial aspects of the nature of gender divisions inside and outside the family. He asserts that perception is one important parameter in the determination of intra-family divisions and inequalities. The informational base of co-operative conflict must include and recognize two biases in *perception*. The first perception bias concerns assessments of interests and well-being that are imprecise and ambiguous. A person may get a worse deal in the collective solution if his or her perceived interest takes little notice of his or her personal well-being. Thus, if a woman undervalues herself, her bargaining position will be weaker and she will be likely to accept inferior conditions (Dwyer and Bruce 1988).

The second perception bias concerns perceptions of contribution, which have to be distinguished from actual contributions since they may be asymmetrical. If a person were perceived as making a larger contribution to the overall well-being of the group than their actual contribution, then the collective solution would be more favourable to that person. It is important to note that perception bias tends to relate to the size of direct money-earning, rather than the amount and time and effort expended (or to the role of non-market activities by other members of the family, who indirectly support such activities). As an example of this, Sen cites the disadvantages of women with frequent pregnancy and persistent childbearing. This makes the outcome of co-operative conflicts less favourable to them, through a lower ability to make a perceived contribution to the economic fortunes of the family. Thus rules governing intra-household distribution appear natural and legitimate although they often embody spectacular inequalities, with gendered perceptions tending to support and sustain them (Sen 1990).

Research by feminist economists provides an important addition to this work. For example, in contrast to the 'one pot, one roof' definition of the household mentioned at the outset of this section, a useful one is that of Feldstein (1986). She identifies it as a system of resource allocation between individuals, in which members share some goals, benefits and resources, are independent on some, and in conflict on others. This research has focused on the causes of widespread intra-household gender inequality, attempting to find out the extent to which this is determined by economic factors as against 'exogenously given' non-economic factors, and the capacity of neo-classic as well as Marxist economic analysis to deal with social and economic relations beyond the market place.[6]

The most important policy-related issue raised by this debate concerns the extent to which intra-household inequality relates to systematic differences in the economic bargaining power of different family members, and consequently the extent to which employment enhances women's domestic decision-making power in co-operative conflict.

Policy-makers now recognize that fundamental gender-based asymmetries in access to welfare and productive resources within households often constrain women's capacity to develop or expand income-generating enterprises. Increasing income is identified as important. Not only does it help women increase their power to negotiate over household assets and thereby decrease intra-household inequalities, but also because it will more effectively influence household welfare. A particular concern is to provide women with greater access to paid employment on the basis that this will increase their control over resources. Evidence clearly shows that working women have more control over the allocation of household resources than do non-earning wives. However, the connection between income-earning and power is not as straightforward as many policy-makers assume.

As Sen's work has identified, perceptions of value play a critical role here. Not only do husbands frequently under-value the economic contribution of women's income to the household (Roldan 1988). The normative expectations that their husband will control their income also often influence women. The question as to who defines 'objective' needs and interests, therefore, is itself highly political, as are questions of whether they will be recognized and on what terms. As Hart (1990) has argued, instead of perception bias it may be useful to focus on how domestic consent is produced and maintained and the conditions under which household members are likely to challenge and redefine rules (rather than measuring 'perceived contribution bias' from some 'objective' norm). 'To come to grips with intra-household bargaining and contestation we have to engage directly with questions of ideology and meaning, and recognise how struggles over resources and labour are simultaneously struggles over meaning' (Hart 1990: 25). Thus Whitehead (1984b), in her comparative study of the politics of domestic budgeting, shows how the relative power of husbands and wives cannot simply be read off wages commanded in the labour market, or labour inputs in agricultural production. Incomes earned by women do not necessarily translate into the same kind of power as that of men. Practices, such as the gender division of responsibility for specific consumption needs, render women's income defined as less important and therefore non-commensurate with that of men.

Feminists also recognize that fundamental changes for women cannot be based solely on increased income. Self-esteem plays an important role in women's potential to mobilize external strengths (such as wages, the persuasion of kin, community opinion). In addition, internal constraints (such as improving their health, education and income-earning capacity) are also important in changing their situation (Dwyer and Bruce 1988). For Sen, along with women's involvement in so-called gainful employment, it is the process of politicization – including a political recognition of gender issues – that can itself cause sharp changes in gender perceptions. The critical issue of increasing women's self-perception of their status and personal power has led to a focus on the role that the collective action of women's solidarity groups play. In the confrontation of the persistent reinforcement of inequalities, their activities are vital; this is a theme that will be returned to in Chapter 9.

THE TRIPLE ROLE OF WOMEN

Of the three planning stereotypes, the most problematic is the third, which relates to gender divisions of labour within the household. While the type-casting of women as 'home-makers' is true, this is only one of three roles that they perform. In most low-income Third World households women have a triple role.[7] 'Women's work' includes not only *reproductive* work, the childbearing and rearing responsibilities, required to guarantee the maintenance and reproduction of the labour force. It also includes *productive* work, often as secondary income earners. In rural areas this usually takes the form of agricultural work. In urban areas women frequently work in informal sector enterprises located either in the home or the neighbourhood. Also, women undertake *community managing* work around the provision of items of collective consumption, undertaken in the local community in both urban and rural contexts.

In most Third World societies the stereotype of the man as breadwinner – that is, the male as productive worker – predominates, even when it is not borne out in reality. Invariably, when men perceive themselves to have a role within the household it is as the primary income-earner. This occurs even in those contexts where male 'unemployment' is high and women's productive work actually provides the primary income. In addition, generally men do not have a clearly defined reproductive role. This does not mean empirically that they do not play with their children or help their women partners with domestic activities. Men also undertake community activities but in markedly different ways from women, reflecting a further sexual division of labour. While women have a community managing role based on the provision of items of collective consumption, men have a community leadership role, in which they organize at the formal political level generally within the framework of national politics.

The triple role and feminist debates

The concept of the triple role is not an arbitrary categorization. It derives from the predominantly feminist debates in the extensive literature on gender relations from both the First and Third World, as well as from the research on women in the Third World. This provides the knowledge base for the new tradition of gender planning.[8] This literature is complex and contradictory. Nevertheless, it is universally agreed that the central problematic remains the concept of power and its opposite, oppression, articulated in gender relations in terms of the subordination of women to men. Furthermore, it is the gender divisions of labour that are identified, above all, as embodying and perpetuating female subordination (Barrett 1980; Mackintosh 1981). This phenomenon, more commonly, if inaccurately, is termed the sexual division of labour. The fact that some tasks are allocated predominantly or exclusively to women, and others to men, is persistent in human society. Until recently such divisions were perceived to be rigid and universal. The fact that this is not true is now clearly accepted. Divisions of tasks at any point in time vary from one country to another. As countries undergo economic change and the nature of work changes so does its distribution between men and women.

It is this recognition that has caused feminists to challenge two basic premises of the gender division of labour; first, that it is 'natural'; second, that the division between the male breadwinner and the female home-maker is based on a perceived complementarity of roles for men and women, who are 'different but equal'. Feminists argue that there is no reason why gender should be an organizing principle of the social division of labour, except the physical process of childbearing. It is the penetration of Western capitalism with its historical separation of production and reproduction that has resulted in such an artificial division, and its ideological reinforcement (Mackintosh 1981).

In examining the different roles of women and men, the gender division of labour provides the underlying principle for separating out and differentiating the work men and women do. It also provides the rationale for the difference in value placed on their work. This accounts for the link between the gender division of labour and the subordination of women. Each of women's three roles will now be examined.

Reproductive work

The reproductive role comprises the childbearing/rearing responsibilities and domestic tasks undertaken by women, required to guarantee the maintenance and reproduction of the labour force. It includes not only biological reproduction but also the care and maintenance of the workforce (husband and working children) and the future workforce (infants and school-going children).

Why is it that the reproductive role is naturally considered women's work? The obvious answer lies in the fact that women bear children and that this connects naturally to the reproduction of all human life. There is no reason why this should extend to the nurturing and caring, not only of children but also for adults, if they are sick or aged, through the daily provision of a range of domestic services. This contradiction reflects the diversity of definitions and meanings of reproductive work.

While 'biological reproduction' refers rigidly to the bearing of children, the term 'reproduction of labour' extends further. It includes the care, socialization and maintenance of individuals throughout their lives, to ensure the continuation of society to the next generation (Edholm *et al.* 1977). This daily renewal and regeneration of the labour force has been termed 'daily' or 'physical reproduction' or 'human reproduction'. It is to these processes that the term 'reproductive role' refers throughout this book.[9] It is important to distinguish this term from that of 'social reproduction'. Mackintosh (1981) defines this to include the far broader processes by which the main production relations in society are re-created and perpetuated. These include not only the production and maintenance of the wage labour force but also the reproduction of capital itself.

Feminists argue that the strict division of labour that makes reproductive work women's work is a consequence of capitalism (Scott and Tilly 1982). Clearly, there are many examples of pre-capitalist societies where such rigid divisions did not exist. However, despite the esoteric examples of the flexible child-care arrangements of pre-colonial and 'traditional' societies provided by such anthropologists as Margaret Mead, the extreme drudgery of domestic work experienced by the vast majority of Third World women, whether they are rural peasants in Asian subsistence production or urban slum-dwellers in Latin American petty commodity production, is also only too apparent.[10]

In describing the 'domestication of women', Rogers (1980) links the increasing importance of women's reproductive role in Western economies to the Industrial Revolution. As the modern cash economy became increasingly divorced from subsistence economy, so women lost economic autonomy in their own right as farmers, craftworkers and traders. In this way they increasingly depended on the wages of men. She comments that in such societies gender distinctions are commonly rationalized by beliefs about the central importance of women's role in childbearing, and the imputed operation of 'maternal instinct'. Western male ideology has been of critical importance in emphasizing the exclusive role of the biological mother in nurturing infants and children. This has reinforced the concept of maternal deprivation and promoted the domestic science movement as a vocation justifying the unemployment of women to serve men. Domestic ideology has reinforced the identification of the domestic sphere and the house as the woman's place. Even when they have a waged job outside the home, women's primary occupation is as wife and mother.

A crucial issue relating to women's reproductive work concerns the extent to which it is visible and valued. For despite its actual character, because it is seen as 'natural' work it is somehow also not real 'work' and, therefore, invisible. This is most graphically illustrated around the issue of rest. When men finish work, be it from the farm or the factory, and return home, they are tired. They therefore rest, whether this takes the form of sleeping, drinking with other men or watching TV. In contrast to this, domestic labour has no clear demarcations between work and leisure; caring for young children is without beginning or end. Because reproductive work is not 'real' work, women very rarely rest except at night. Consequently in most societies women tend to work longer hours than men. Not only are they the first to get up to prepare the household for the working day, but also last to go to sleep.

Lack of recognition of the economic cost of reproductive work under capitalism has resulted in the separation of paid work, which is allocated an exchange value, from that of unpaid 'domestic' work, which is allocated only a use-value (CSE 1976). The fact that this occurs not only in Western but also in non-waged economies, however, suggests that capitalism is not the only explanation. In both capitalist and socialist societies, men do not have clearly defined reproductive roles. In contrast, women's allocation of domestic work, particularly child care, remains extraordinarily rigid and persistent at a global level. Why is it that the gender divisions of labour around human reproduction are so rigid? This question requires an understanding of the relations under which reproductive work is done. Marriage-based households are constructed by definition on the basis of gender, with economic relations within such households also structured by gender (Whitehead 1981). House-work and child care are the activities most influenced by the relations of marriage. It is this that provides the critical link between productive and reproductive work. As Mackintosh so succinctly argues,

> The household has become a kind of mediating institution, mediating that is two sets of social relations; that of marriage and filiation, which act to constitute the household and determine the context of much of child-care, and the wider economic relations of society. Women's performance of domestic work, especially the care of children within the home, both expresses their dependence and subordination within marriage (since men actively benefit from this work) and also weakens their position within the labour market, contributing to their low wages and poor conditions as wage workers.
>
> (1981: 11)

Productive work

The productive role comprises work done by both women and men for payment in cash or kind. It includes both market production with an exchange value, and subsistence/home production with an actual use-value, but also a potential exchange value. For women in agricultural production this in-cludes work as independent farmers, peasants' wives and wage workers.

Whereas the ideology of patriarchy has served to reinforce the popular stereotype of the male breadwinner, reality does not bear this out. Throughout the Third World most low-income women have an important productive role. Nevertheless, the rigidity of gender divisions of labour has ensured that although this is the one area in which both men and women work, they do so unequally. Ideologically masked asymmetrical gender relations in product-ive work, whether it is in the formal or informal sector, rural or urban production, means that again women as a category are subordinated to men.

Definitions of 'productive' work are fraught with complexities. In this book productive work is broadly defined as a task or activity which generates an income and, therefore, has an exchange value, either actual or potential. This is most visible in cash economies, and by contrast, least so in subsistence production. It includes work in both the formal and informal sectors, as well as in family enterprises. In the latter case it may not be perceived of as work with an exchange value, since work undertaken does not get a wage.

Many feminists have argued that since the reproduction of labour involves several productive tasks, a dualist division, such as that between production and reproduction, is not helpful. They claim that reproductive work is also productive. Because its domain is the production of use-values under non-wage relations, it is not identified as 'productive' work (see, for instance, CSE 1976; Gardiner 1977; and Barrett 1980). It is recognized that the use of the term 'productive' work as referring only to work with an exchange value is an over-simplification of reality. Nevertheless, in developing the conceptual principles for gender planning the purpose of distinguishing between women's productive and reproductive roles is precisely to highlight the multiple forms of women's work. This reveals the severe limitations of contemporary planning categories, which in emphasizing the difference between men's productive work and women's reproductive work, have so effectively rendered women's productive work invisible. The purpose of this simplification, therefore, is not to under-value or ignore the importance of production for use-value. It is to identify how the complexities of women's multiple interdependent roles can be simplified into methodological tools that planners can translate into practice.

The last decade has witnessed a veritable explosion of research on the complexities of women's productive work.[11] This has clearly identified the extent to which gender divisions of labour continue to reinforce women's subordinate position in productive work. Despite structural transformations in systems of production in both urban and rural areas, this continues. Although patterns of segregation run through all societies, exactly which job falls to men, and which to women, has varied enormously. In essence, however, it is still the fact that 'women pick up the work men won't do'.[12] This is most visible where men and women both work for wages. Segregation of the labour market means that in all economies women predominate at the lower end of the labour market. They cluster in certain industrial and agricultural sectors, and within these tend to have certain occupations. Not only are they distributed vertically, that is to say, sex-segregated based on gender hierarchy, into lower-paid and lower-skilled jobs. They are also distributed horizontally within particular sectors, with few women in managerial positions and most in those occupations that are an extension of domestic labour (Moser 1981).[13]

Even where entirely new areas of employment are created, new forms of 'women's work' are established, thereby perpetuating women's subordination. This is most visible in the processes of the internationalization of capital, when production relocates or subcontracts to avoid trade-union regulations in Free Trade or Border Zones. It takes advantage of lower employment costs with matching or even higher levels of productivity. Most frequently this is young, single women's labour. Elson and Pearson (1981) have identified that such women are often recruited because of their so-called 'nimble fingers'. In reality, they form the cheapest, most docile labour force best suited for tedious, monotonous waged work, whose lower wages are attributed to their secondary status in the labour market, which is itself identified because of their capacity to bear children.

Neither is the situation generally any better for women outside waged work, whether self-employed or working as members of family enterprises. Small scale 'petty commodity production' within the so-called informal sector is both dependent and involutionary in nature. Its relationship to large-scale formal sector production is exploitative in terms of issues such as access to credit, markets and raw materials (Moser 1978, 1984; Schmitz 1982). However, for women working in this sector, exploitation by the formal sector is not the only constraint they confront. Women encounter additional constraints in their gendered role. This occurs particularly in household enterprises where men recruit their wives to work unpaid in tasks such as sewing, weaving or cooking (Goddard 1981; Moser 1981; Jumani 1987). The fact that this work is an extension of reproductive domestic work into the market means that often both men and women fail to make the distinction between remunerated and non-remunerated work, and therefore do not perceive it as work.

Gender divisions of labour also continue to structure work relations in rural areas. In agricultural production in Africa the common policy stereotype is still the dichotomy popularized by Boserup's seminal research; women work on subsistence food production while men produce cash crops. This results in a high level of invisibility of rural women's work (Beneria 1979, 1982; Dixon-Mueller 1985). As Whitehead (1990) has identified, the gender divisions of labour are more complex. Although there are clear gender demarcations of tasks, women generally have a dual productive role, and sometimes even threefold. In cases where women have separate access to land it is common for them to work both as 'independent farmers' on their own smallholder plots, and work as 'peasant wives', contributing to household production as unremunerated labour in the fields of male household members, where they work in planting, hoeing and weeding, the tasks designated in the gender division of labour as women's work. Women also work as wage labourers, most frequently seasonally, to supplement household income. Finally, the commercialization of agriculture and concomitant transformations in production systems and technology have influenced the so-called 'separate but equal' status of women in several fundamental ways. These relate particularly to time and access to resources. Capitalist agriculture has put increasing pressure on women to spend more time working on their male kin's farm. This reduces the income, in cash or kind, from a woman's own smallholding. The change of ownership from collective to market systems of private ownership, with state-codified individual forms of land allocation and resettlement, has frequently resulted in blindness to, or ignorance of, women's land rights (Dey 1981; Muntemba 1982; Palmer 1985).

Community managing and community politics

The community managing role comprises activities undertaken primarily by women at the community level, as an extension of their reproductive role. This is to ensure the provision and maintenance of scarce resources of collective consumption, such as water, health care and education. It is voluntary unpaid work, undertaken in 'free time'. The community politics role in contrast comprises activities undertaken by men at the community level organizing at the formal political level. It is usually paid work, either directly or indirectly, through wages or increases in status and power.

Recognition that women have a community managing role is still far from widespread, such that it is still most frequently identified as part of reproductive work. This may well relate to the fact that feminist debates around issues of consumption are preliminary. In outlining the conceptual principles of gender planning, however, it is essential to define this as a clearly identified role in its own right. For, like reproductive work, community managing is seen as 'naturally' women's work. The importance of giving recognition and visibility to this form of work, as an activity in its own right, is of particular significance in the current economic climate where low-income households are increasingly resolving community-level problems through self-help solutions.

'Community managing', then, is defined as the work undertaken at the community level, around the allocation, provisioning and managing of items of collective consumption.[14] A wide diversity of consumption needs for the reproduction of labour power have been increasingly socialized at the level of the community. The fact that women, in their acceptance of the gender division of labour, see the house as their sphere of dominance and take primary responsibility for the provision of consumption needs within the family has already been discussed. However, these needs include not only individual consumption needs within the household, but also the consumption needs of a collective nature at the neighbourhood or community level. Thus, for women the point of residence includes not only the home but also extends into the surrounding areas. Social relationships include not only household members but also neighbours. Mobilization and organization at the community level is a natural extension of their domestic work. Where there is open confrontation between community-level organizations and state or NGOs for infrastructure provision, women most frequently take primary responsibility for the formation, organization and success of local-level protest groups. This phenomenon has been most widely documented in urban South America (see Barrig and Fort 1987; and Moser 1987b). It is, however, neither uniquely South American nor urban. Recent examples – for instance, by Barrett *et al.* 1985; Sharma *et al.* 1985; Shiva 1988; Yoon 1985; and Omvedt 1986 – have illustrated the community managing role of women in rural environmental and basic service struggles in Asia and Africa.

It is now widely recognized that the serious deficit in housing, infrastructure and social service provision experienced in most Third World countries is not simply a problem of rapid urbanization. In the late 1970s Castells (1977) described these gaps in provision as a 'crisis of collective consumption'. He argued that while the capitalist system requires the adequate reproduction of labour power as a prerequisite of continued production and accumulation, private capital finds it less and less profitable to invest in the means of collective consumption. In order to deal with this contradiction the state intervenes (directly or indirectly) and takes responsibility for the reproduction of labour power.[15] This may have provided an interpretation for increased top-down state provision in the 1970s through such policies as basic needs. A decade later debt, recession and structural adjustment loans have resulted in severe cutbacks in state provision of infrastructure and services. For low-income women this has meant increased time pressure in their role as community managers. Here they engage in bottom-up struggles manifested through self-help community-based solutions to obtain food, health and education. In contexts where NGOs with highly 'participatory' programmes are helping with service delivery these are most frequently designed on the assumption that women will provide the necessary (unpaid) labour. This exacerbates the situation even more – an issue that will be further examined in Chapter 9.

It is important to note that men also work at the community level. However, gender divisions of labour are as important here as they are at the household level. The spatial division between the public world of men, and the private world of women, means that for women the neighbourhood is an extension of the domestic arena, while for men it is the public world of politics. This means that while women in their gender-ascribed roles of wives and mothers are involved in community managing, men are involved in community politics. In low-income communities throughout the world there is a consistent trend for political organizations to be run by men with mainly male members, and for collective consumption groups to be in the hands of women. For example, in Lima men control and lead the Junta Communal, while women organize the Community Kitchen Associations; in Manila a man is generally the Baranguay Captain, while a woman obviously leads the Women's Club; in Bombay, the National Slum Dwellers Association local representative is a man, but the Mahila Mandal leader is a woman (Moser 1987c).

In organizations in which these two activities overlap, especially in societies where men and women can work alongside each other, women most frequently make up the rank-and-file voluntary membership. Men tend to be involved in positions of direct authority and often work in a paid capacity. The fact that male leaders are frequently paid for their work is legitimized by the fact that 'a man has to work'. Women, by contrast, are expected to be selfless and 'pure'. Their participation is justified in their gender-ascribed role of being a good mother, working to improve living conditions for their families (Moser 1987b). This gender division at the community level between paid men's work and unpaid women's voluntary work has been extensively reinforced by governments, international agencies and NGOs alike. For example, urban basic services programmes, such as those of UNICEF in India, are often designed to provide paid employment for men in official positions. Their successful implementation also requires the unpaid work of women in the community. Equally, when rural water-pump maintenance programmes relying on local government employees failed, they were redesigned to use local women to maintain the pumps, but in an unpaid capacity.

As long as women mobilize around issues relating directly to their social sphere and outside established political organizations, they can become very powerful, precisely because they do not challenge the nature of their gender subordination. Once they move, however, into the 'masculine' world of public politics confrontations are virtually inevitable. These are personal, with women friends and husbands, as well as political, with male politicians; women only avoid conflict by rigidly conforming to their gender-ascribed roles.[16] This may be one important reason why women with their particular responsibilities for social and welfare politics often choose to remain involved in community managing. For, as Kaplan has commented, 'The bedrock of women's consciousness is the need to preserve life' (1982: 546).

3   Practical and strategic gender needs and the role of the state

An important underlying rationale of gender planning concerns the fact that men and women not only play different roles in society, with distinct levels of control over resources, but that they therefore often have different needs. This chapter provides a description of the concept of gender interests and its translation into planning terms as gender needs. It identifies the important distinction between practical and strategic gender needs. A brief description follows of the way the state in different political contexts effectively controls women's strategic gender needs through family policy relating to domestic violence, reproductive rights, legal status and welfare policy. The usefulness of these gender planning tools is then examined in terms of several interventions in different planning sectors.

At the outset it is important to emphasize that the rationale for gender planning does not ignore other important issues such as race, ethnicity and class. It focuses specifically on gender precisely because this tends to be subsumed within class in so much of policy and planning.

THE IDENTIFICATION OF GENDER NEEDS

Planning for low-income women in the Third World must be based on their interests – in other words, their prioritized concerns. When identifying interests it is useful to differentiate between 'women's interests', strategic gender interests and practical gender interests, following the threefold conceptualization made by Maxine Molyneux (1985a). Having identified the different interests of women it is possible to translate them into planning needs; in other words, the means by which their concerns may be satisfied.[1]

From a planning perspective this separation is essential because of its focus on the planning process whereby an *interest*, defined here as a 'prioritized concern', translates into a *need*. This in turn is defined as the 'means by which concerns are satisfied'. A further distinction can then be made concerning women's needs, strategic gender needs and practical gender needs. With this distinction, gender policy and planning can be formulated, and the tools and techniques for implementing them clarified. For example, if the strategic gender interest – namely, the prioritized concern – is for a more equal society, then a strategic gender need – that is, the means by which the concern may be satisfied – can be identified as the abolition of the gender division of labour. On the other hand, if the practical gender interest is for human survival, then a practical gender need could be the provision of water.

Women's interests and gender interests

At the outset it is important to clarify the distinction Molyneux makes between 'women's interests' and gender interests. The concept of 'women's interests' assumes compatibility of interests based on biological similarities. In reality the position of women in society depends on a variety of different criteria, such as class and ethnicity as well as gender. Consequently, the interests they have in common may be determined as much by their class position or their ethnic identity as by their biological similarity as women. As Molyneux (1985a) has argued, women may have general interests in common. But this should be referred to as 'gender interests', to differentiate them from the false homogeneity imposed by the notion of 'women's interests':

> Gender interests are those that women (or men for that matter) may develop by virtue of their social positioning through gender attributes. Gender interests can be either strategic or practical each being derived in a different way and each involving differing implications for women's subjectivity.
>
> (1985a: 232)

Similarly, within the planning context 'women's needs' also vary widely. They are determined not only by specific socio-economic contexts, but also by the particular class, ethnic and religious structures of individual societies. Consequently, although planners refer to the category of 'women's needs' in general policy terms, it is of limited utility when translated into specific planning interventions.

Gender needs

Molyneux's distinction between strategic and practical gender interests is of theoretical significance for gender analysis. For gender planning it is the distinction between strategic and practical gender needs that is important. It is this that provides gender planning with one of its most fundamental planning tools. Frequently, different needs are confused. Clarification is essential if realistic parameters are to be identified both as to what can be accomplished in the planning process, as well as the limitations of different policy interventions.

Strategic gender needs

Strategic gender needs are the needs women identify because of their subordinate position to men in their society. Strategic gender needs vary according to particular contexts. They relate to gender divisions of labour, power and control and may include such issues as legal rights, domestic violence, equal wages and women's control over their bodies. Meeting strategic gender needs helps women to achieve greater equality. It also changes existing roles and therefore challenges women's subordinate position.

Strategic gender needs are those needs that are formulated from the analysis of women's subordination to men. Deriving out of this analysis strategic gender interests necessary for an alternative, more equal and satisfactory organization of society than that which exists at present can be identified. This relates both to the structure and nature of relationships between men and women. As will be illustrated, the strategic gender needs identified to overcome women's subordination vary depending on the particular cultural and socio-political context within which they are formulated. Strategic gender needs, as Molyneux has identified, may include all or some of the following:

> The abolition of the sexual division of labour; the alleviation of the burden of domestic labour and childcare; the removal of institutionalized forms of discrimination such as rights to own land or property, or access to credit; the establishment of political equality; freedom of choice over childbearing; and the adoption of adequate measures against male violence and control over women.
>
> (1985a: 233)

Strategic gender needs such as these are often identified as 'feminist', as is the level of consciousness required to struggle effectively for them. Historically, top-down state intervention alone has not removed any of the persistent causes of gender inequality within society. The capacity to confront the nature of gender inequality and women's subordination has only been fulfilled when it has incorporated the bottom-up struggle of women's organizations, as will be further discussed in Chapter 9. Even here, however, despite a few optimistic examples, the failure to fulfil strategic gender needs continues to be a widespread preoccupation for many. As Molyneux has identified, for feminists it is these which are women's 'real' interests (Molyneux 1985a).

Practical gender needs

Practical gender needs are the needs women identify in their socially accepted roles in society. Practical gender needs do not challenge the gender divisions of labour or women's subordinate position in society, although rising out of them. Practical gender needs are a response to immediate perceived necessity, identified within a specific context. They are practical in nature and often are concerned with inadequacies in living conditions such as water provision, health care, and employment.

Practical gender needs, in contrast, are those that are formulated from the concrete conditions women experience. These derive from their positions within the gender division of labour, in addition to their practical gender interests for human survival. Unlike strategic gender needs they are formulated directly by women in these positions, rather than through external interventions. Practical needs, therefore, are usually a response to an immediate perceived necessity which is identified by women within a specific context. As Molyneux has written, 'they do not generally entail a strategic goal such as women's emancipation or gender equality... nor do they challenge the prevailing forms of subordination even though they arise directly out of them' (1985a: 233).

The gender division of labour within the household gives women primary responsibility not only for domestic work involving child care, family health and food provision, but also for the community managing of housing and basic services, along with the capacity to earn an income through productive work. Therefore, in planning terms, policies to meet practical gender needs have to focus on the domestic arena, on income-earning activities, and also on community-level requirements of housing and basic services. In reality, basic needs such as food, shelter and water are required by all the family, particularly children. Yet they are identified specifically as the practical gender needs of women, not only by policy-makers concerned to achieve developmental objectives, but also by women themselves. Both are, therefore, often responsible for preserving and reinforcing (even if unconsciously) the gender division of labour. Since there is often a unity of purpose between the development priorities of intervening agencies and practical gender needs identified at the local level, the two frequently and easily become conflated. This serves the purposes of planners who are then identified as meeting 'women's needs'. At the same time it can make it even more difficult for women themselves to recognize and formulate their strategic gender needs.

It has become very popular for policy-makers and the media alike to label any policy or programme associated with women as 'feminist' or 'women's lib'. Such terms are used by many in such a derisory manner that they provoke a hostile and negative reaction from female and male planners alike. The differentiation between practical and strategic gender needs provides a critical planning tool. This allows practitioners to understand better that planning for the needs of low-income women is not necessarily 'feminist' in content. Indeed, the vast majority of interventions for women world-wide are concerned with them within the existing gender division of labour, as wives and mothers. These are intended to meet their practical gender needs. While such interventions are important, they will only become 'feminist' in content, if, and when, they are transformed into strategic gender needs.

Gender needs differentiation, therefore, can provide a useful tool for planners. Not only does it help in diffusing the criticisms of those who find 'feminism' unacceptable by showing them that working with women is often not 'feminist'. In addition, it is helpful for policy-makers and planners responsible for meeting the practical gender needs of women, in assisting their adoption of more 'challenging' solutions. What then defines the 'political space' for addressing different gender needs within specific contexts? Is it the family, civil society or the state? To answer such a question requires an understanding of the interrelationship of these three levels of social, economic and political organization.

THE STATE AND ITS CONTROL OVER WOMEN'S STRATEGIC GENDER NEEDS

The extent to which the state mediates relationships not only between state and civil society, but also those within the family between men, women and children, has critical implications for the identification of the 'room for manoeuvre' to address strategic gender needs. The last decade has witnessed an important expansion of research on the state and the extent to which it liberates or controls the lives of women through a diversity of social, economic, political and legal policies. In different contexts, marriage laws, legal provisions regarding rape and abortion, and population-control policies control female sexuality and fertility. Similarly, state laws governing wages, taxation and social security benefits have combined to reproduce gender divisions of labour within the family. In the space available a complex topic such as this cannot be reviewed comprehensively. The intention is to provide some very general indications as to the implications in terms of meeting both practical and strategic gender needs.[2]

Does the state control women?

Just as there are different interpretations of the origins and dynamics of female subordination, so too there are different feminist analyses of the state, and the extent to which it controls women. Radical feminists, for instance, identify sexual politics as the central area of struggle, and traditionally have stressed the biological basis of the sexual imbalance of power, thereby often being identified as promoting biological determinism (Hartmann 1981). They see the basic motivating force of history as men's striving for domination and power over women, with the physical subjugation of women by men as the most basic form of oppression (Jagger 1977). Radical feminists emphasize the importance of patriarchy in reinforcing women's subordination, which they define as the system of sexual hierarchy in which men possess superior power and economic privilege. Patriarchy is maintained through male control over such arenas of power as politics, industry, religion and the military both within and outside the state.

Marxist feminists, in contrast, see the capitalist state as the root of women's oppression. Capitalism is particularly responsible for women's double oppression in productive and reproductive work (Barrett 1980); that women commonly form the cheapest, most vulnerable part of the labour force is in the interests of capitalism; that women, who through their domestic labour, provide services which would otherwise have to be paid for, lower the cost of real wages, is also in the interest of the state, since it would otherwise have to pay to provide the services necessary to ensure the reproduction of the labour force for capitalism. If capitalism is viewed as oppressing women, has socialism liberated them?[3] It was Engels who first identified women's particular subordination as being derived from their position within the monogamous family. He argued that their emancipation from patriarchy and control would come with their entrance into productive work, with their domestic reproductive work being taken over by the state. This view has dominated thinking in socialist economies where it was envisaged that women would join men in the proletarian struggle to overthrow capitalism (Hartmann 1981: 4).

Writing in 1981, Molyneux (1981: 1–2) identified three main areas where the aims of capitalist and socialist states in principle do differ. First, socialist states officially recognize women's oppression and identify it as a social and political problem requiring official intervention, typically in the form of strategic gender needs met top-down by the state. This contrasts with capitalist states, where historically even enfranchisement has been a struggle (1981: 8); secondly, under socialism, labour-force participation is a normal expectation for every active woman, if not a duty to both the state and their families. Socialist states try to maintain high employment levels (thereby meeting practical gender needs for gainful employment through top-down intervention). By contrast, in capitalist states women are more likely to form a large part of the reserve army; thirdly, socialist states accept a greater responsibility for the reproduction of labour power and for social welfare, particularly in the support structures for a diverse number of practical gender needs relating to child care, health education and community-level services. Although many such provisions may have been insufficient, the degree of intervention is far greater than in capitalist economies with a comparable level of economic development.

In many socialist states legal reform has been used to challenge traditional practices; through legislation they have introduced top-down strategic gender needs, redefining family relations between both men and women and parents and children. These have included the banning of polygamy, child marriage and marriage payments. However, despite the existence of legislation with the potential to meet strategic gender needs, men have continued to subordinate women under socialist systems. Women continue to cluster in less-skilled, lower-paid and lower-status jobs. They fail to reach positions of power, and commonly undertake most of the domestic work, or where this has been socialized, it remains under-valued 'women's work'. The socialist feminists, who emerged to deal with the 'unhappy marriage' of Marxism and feminism, therefore recognized the pervasiveness and persistence of patriarchy within and across societies and classes even in socialist countries (Maguire 1984; Sargent 1981).

Why is it that the gender divisions of labour around human reproduction are so rigid, regardless of state intervention? Socialist feminists argue that this is because the gender allocation of tasks is most crucial for the perpetuation of existing relations of marriage, procreation and filiation. It is within these that male dominance and control over female sexuality is incorporated. Consequently, it is around human reproduction that gender division of labour is identified most strongly as 'natural'. Any changes are perceived as a serious threat to established forms of masculine and feminine gender identity. Thus the perpetuation of women's subordination lies in the interrelationship between their position in the marriage-based household and in the wage sphere. Women work not only for capital but they also work directly for men (Maguire 1984). Therefore men have vested interests in women's subordination. In some contexts, both men and women may have wanted to change the exploitative conditions of capitalism, but they may not be struggling for the same transformations. As is only too well known, national liberation has not necessarily meant women's liberation.[4]

The 1990s has already witnessed the destruction of many socialist and communist state systems as well as the growing influence of religious fundamentalism as the basis for state legitimacy. Hence, the simple dichotomy between capitalism and socialism is no longer as useful an analytical tool as it seemed a decade ago. This makes it even more important to deconstruct state policy as it relates to different aspects of women's lives.

Where is state policy towards women located?

If feminists cannot agree on the degree to which the state oppresses women, they concur that it is the *family* that is 'the core site of women's oppression' (Pascall 1986: 36). For it is the family that is identified as providing the boundary between the public and the private domains; that is, the boundary of state interference in individual existence. Consequently, it is in policy towards the family in which the state intervenes most powerfully in the lives of women. This applies both to legislation and planning interventions, and to countries with very different 'political' systems. For as Molyneux (1985b) has identified, even socialist states did not destroy the family. They sought through their constitutions to create a special family form as the basic unit of society to ensure productive and social stability. State policy towards women within the family, however, is highly complex, because it is not unidirectional. It varies depending on its purpose, alternatively controlling and supporting women.

a) Domestic violence

At one level the family is seen as a *private* domain in which the state only interferes in special circumstances. Here state policy concerns the articulation of relationships between the public and private spheres. The concept that family matters are a private business gives husbands licence to treat wives as they wish behind closed doors. For women this has been particularly problematic over such issues as domestic violence and sexual abuse. Violence is closely linked to unequal family structures and women's economic dependence on men. In addition, social stigma attached to single women frequently forces women to stay married. Writing of the dilemma of a Mozambique woman, Urdang (1989) recalls, 'she finds it too difficult to go against social attitudes, she is afraid that if she is alone she will be labelled a prostitute by the community' (p. 179). Because domestic violence occurs in the private sphere of the home, it is most frequently invisible. Even when domestic violence is identified, relatives, as much as judges, the courts or the police, are reluctant to interfere. Many states identify such strategic gender needs as control over their sexuality and bodies (and consequently also lives) as a 'private' family matter between man and wife, and outside their mandate. Even when laws exist on the statute books, police lethargy in its enforcement is common. In 1979 a women's campaign occurred in Delhi against dowry murders – bride burning of women who had not paid full dowry to their husbands' families. Kumar (1989) explains that this was 'the first time the private sphere of the family was invaded, and held to be the major site of the oppression of women' (p. 20).[5]

b) Female fertility and women's reproductive rights

One of the best-known 'special circumstances' in which the state is quite willing to intervene inside the family relates to women's reproductive rights, including its control over female fertility. Such policies are justified in terms of a number of different reasons. Anti-natalist policies to limit population, in countries such as China (Davin 1987) and India (Editorial Collective 1987), relate to economic growth priorities and the fear that the 'population explosion' could offset this; pro-natalist policies can relate to demands for labour in the economy, as has happened in Eastern Europe and the then Soviet Union (Molyneux 1985b). Nationalism and the concern with the survival or expansion of a particular group, based on ethnic, religious or other criteria, in countries as diverse as Malaysia (Stivens 1987) and Israel, have also resulted in pro-natalist policies. As Yuval-Davis (1987) has described, 'the most important national role of Israeli Jewish women relates to the army, namely as reproducers' (p. 199).

For women, the problem with population policy, whether pro- or anti-natalist, is that it fails to recognize their strategic gender need to control their own fertility. If the birth-rate is left to women, it obviously gives them great power. The right to choice is commonly equated with the right to have children or not to have children. While access to education and employment has decreased female fertility, it is clear that men and women have different interests in children. Women may continue to want more children for both economic and social reasons, relating to the power, status and prestige that comes from motherhood.

In its control over women's reproduction, state intervention varies from attractive incentives to drastic disincentives. Nowhere is the latter more explicit than in China's one-child family policy, which is also unusual in targeting both parents as responsible. Disincentives for more than one child include fines, salary deductions and job promotion limitations imposed on both parents. Nevertheless, the main burden of such policy falls on women in terms of both health effects of multiple abortions and sterilizations. In addition, female infanticide increases as parents keep struggling to have sons. This suggests that top-down policy has not changed the traditional devaluation of daughters and preference for sons (Davin 1987). A similar problem has emerged in India. An active family-planning policy – at times even coersive – in urban areas has resulted in increased amniocentesis (testing for foetal abnormalities) and the aborting of female foetuses. The extent to which female children are devalued is illustrated by a 1982 survey by the Women's Centre in Bombay, which found that 7,999 out of 8,000 abortions were of female foetuses (Mathai 1990).

c) The legal status of women in marriage

In other contexts, the state views the family as *fragile and needing support*, with the latter's 'collapse' identified in terms of the 'breakdown' of the 'fabric' of society. State morality can provide the basis for legal control over women's status within the family. The conflation of female sexuality and social order has important implications for women's strategic gender needs relating to control over their lives.

Where religious oppression is institutionalized by the state, as in Iran, this is often manifest in an extreme form. Afshar (1987) maintains that in Iran it is the interpretation of Islam by male religious leaders that is at the root of women's oppression. The Koran itself guarantees certain economic and religious rights to women, even if defining them only as mothers and inferior to men. The religious fundamentalist revival in the post-revolution state of Khomeini after 1979 identified 'women's liberation' as yet another export of Western imperialism. This resulted in a severe reduction in women's rights, which include the following; men and women have unequal citizen status, such that women's evidence as witnesses in their own right is not accepted unless corroborated by a man; women's freedom of dress and movement is curtailed with the compulsory reinstatement of the *hejab* (veil) outside the home, based on the ideology that women are a source of corruption and sexual provocation to men, and therefore must be covered; with the revoking of the 1967 and 1976 Family Protection Acts polygamy of up to four wives is legally permissible, based on the argument that this is a mechansim to prevent women's destitution or prostitution. Afshar argues that in practice polygamy does not protect women, but merely displaces them to make room for other potential mothers, such that 'the husband has become an absolute ruler, entitled to exercise the power of life and death in his home' (p. 83).

It is important to realize that even in countries with highly sophisticated civil legal systems, the coexistence of religious and civil law can effectively oppress women. The constitution of India guarantees equality for all, outlaws discrimination on the grounds of race, caste and birth, and promotes equality of opportunity in employment for all. But, despite constitutional rights of all citizens under civil law, religious laws for each religious group have jurisdiction over all family matters such as marriage, divorce, child custody and dowry. A single civil law pertaining to family matters does not exist. The reason is the 'communalist' agreement that such a uniform code would be an interference with the diverse rights of some eight religious groups with their own laws to determine their own personal matters. However, by also allowing religious autonomy, religious laws can deny women these rights, as well as preventing them from claiming rights under civil law.[6]

d) Benefit rights and welfare policy

Finally, the state sees the family as the main organizational structure through which it can reinforce its control over differences in the benefit status of individuals, most specifically in its definition of men as independants and women as dependants. This is not a new phenomenon. Historically, European patriarchal ideology influenced policy towards the family. The separation of public and private spheres reinforced the model of the dependent family with the male 'breadwinner' and female carer, which was then crystallized in legislation and welfare policy (McIntosh 1979). A classic example is provided by the social reformer, Beveridge. In his vision about the post-World War II welfare state, he was unequivocal about women's role. 'In the next thirty years housewives as mothers have vital work to do in ensuring the adequate continuance of the British race and of British ideals in the world' (Beveridge 1942: 53). The UK welfare-state legislation is still salutary today, given the widespread export of British welfare policy to its ex-colonies.[7]

Support of a particular family form meant that the welfare state sustained rather than alleviated the dependency of women on men within marriage through a diversity of policies. As Pascall (1986) has detailed, this status resulted in a separate insurance class for 'housewives, that is married women of working age' (p. 10), most of whom it was assumed would 'make marriage their sole occupation' (p. 49). Beveridge concluded that the married woman's benefits need not be 'on the same scale as the solitary woman, because among other things her home is provided for her' (p. 50). Thus, married women would have 'contributions made by the husband' (p.11). Simultaneously the welfare state failed to reduce women's dependency within the family by recognizing reproductive work through redistributive income and financial maintenance.

State control of women also extends beyond the family to control of their position within the labour market. Here the relationship between reproduction and production is at its most explicit with the historic concept of the legally enforced 'family wage'; that is, of a man's wage being sufficient to support two adults and their children. Even where equal pay legislation promises equal pay for women, their wages generally have continued to be depressed. Low pay does not meet strategic gender needs for independent control over resources, but interacts with, and reinforces, women's economic dependence within the family.

MEETING PRACTICAL GENDER NEEDS IN PLANNING PRACTICE: DOES IT PROVIDE AN ENTRY POINT FOR MEETING STRATEGIC GENDER NEEDS?

Given the very real interests of the state, civil society and men in subordinating women through control of their status, bodies and indeed sometimes even their lives, there are widespread constraints in meeting strategic gender needs. Consequently, planners often seek to use practical gender needs as an entry point for more fundamental change. The following examples of interventions in such sectors as employment, housing, rural production, the environment and basic urban services illustrate the potential and limitations of different planning practices to reach both sets of needs within specific planning contexts. Some of these examples are graphically illustrated in Table 3.1.

However, even in planning for practical gender needs it is necessary to recognize that women require integrative cross-sectoral planning strategies. For instance, most governments base national planning priorities on a sectoral approach. Employment planning, for instance, is concerned with individuals as paid workers. It assumes a household support system, while women's participation in the labour force is constrained by their triple commitment. Social-welfare planning which concentrates on the child-rearing roles of women, does not adequately take account of their income-earning activities. For example, health facilities in low-income areas are frequently under-subscribed because their opening hours are inappropriate for working mothers. The failure to recognize women's multiple responsibilities may not merely jeopardize the implementation of policy, with programmes frustrating rather than meeting basic needs; perversely, it may in fact worsen the position of women.

Gender needs in employment

With the lives of the majority of low-income women dominated by the necessity to generate an income, one fundamental problem faced is the lack of adequate skills. The provision of training, therefore, meets an important practical gender need allowing access to employment. How far it also reaches more strategic gender needs depends not only on whether it increases women's economic independence, but also on the type of training. In many ex-British colonies community development centres for decades have provided women's training courses in home economics, introducing skills

Table 3.1 Women's triple role and practical and strategic gender needs

Type of intervention	Women's role recognized			Gender need met	
	R	P	CM	PGN	SGN
1 *Employment policy*					
a) Skill training					
Cooking angel cakes	X			X?	
Dressmaking		X		X	
Masonry/carpentry		X		X	X(a)
b) Access to credit					
Allocated to household		X		X	
Allocated to women		X		X	X(b)
2 *Human settlement policy*					
a) Zoning legislation					
Separates residence and work	X				
Does not separate residence and work	X	X		X	
b) House ownership					
In man's name	X			X	
In woman's name	X	X		X	X(c)
3 *Basic services*					
a) Location of nursery					
Located in community	X	X	X	X	
Mother's workplace	X	X		X	
Father's workplace	X	X		X	X(d)
b) Transport services					
Only peak hours bus service		X		X?	
Adequate off-peak service	X	X	X	X	
c) Timing of rural extension meetings					
In the morning		X		X	
In the afternoon/ evening	X	X	X	X	

R = Reproductive   P = Productive   CM = Community managing
PGN = Practical gender need   SGN = Strategic gender need
(a) Changing the gender division of labour
(b) Control over financial services
(c) Overcoming discrimination against women owning land, by law or tradition
(d) Alleviation of the burden of domestic labour

intended to assist women become better providers within the household. Such training recognizes the reproductive role of women, and can meet practical gender needs relating to basic health and nutrition but does not recognize women's productive role and the important practical gender need to earn an income (see Table 3.1).

In a recent example from Nigeria, the curriculum in a community development home economic course included new recipes such as angel cakes to be cooked in Western-style ovens (Erinle 1986). The majority of low-income women dropped out of the programme, preferring to cook 'traditional' *gari* (processed cassava) food for sale. The training recognized the reproductive role of women and might have met nutritional needs, assuming that angel cake had greater nutritional value than the traditional foods for which it substituted. But it failed to recognize the productive role of women, which was the priority for the women themselves.

By contrast, skill training in primary school teaching, nursing and dressmaking often meets income-generating needs. A common type of training is dressmaking, taught throughout the world, at a range of levels from government programmes, through medium-sized NGO projects, to small, self-help groups. In diverse cultures and contexts the underlying rationale for the provision of dressmaking skills is similar: that this is a skill women already know, or should know, and one they can use not only in the home but also to earn an income. Such training can meet a practical gender need. Because dressmaking is an area in which women traditionally work, this does not challenge the gender division of labour.

The training of women in areas traditionally identified as 'men's work' may not only widen employment opportunities for women, but may also break down existing occupational segregation, thereby contributing to the strategic gender need to abolish the gender division of labour. Women's training in house-building skills such as masonry and carpentry provides one such example. Although in most societies women traditionally are involved in rural house building, the urban-based development of a female formally skilled labour force for house construction has been prevented by an occupational sex segregation, with construction designated as 'men's work', other than in those contexts, such as India, where women provide unskilled labour (NIUA 1982). Skill training for women in the construction sector often meets with hostility and resistance, precisely because it challenges the existing gender division of labour. As case studies from Jamaica (Schmink 1984a) and Nicaragua (Vance 1987) show, training and the tacit acceptance of male colleagues both assist women construction workers to find work either in existing projects or in the construction sector generally.

For rural women involved in agricultural production and processing, access to credit is often tightly constrained. Frequently, even in credit schemes designed to assist women, there is a tendency to allocate resources at the family or household level, despite the fact that they generally remain in the hands of the male household head. A typical example is a project to provide loans to landless families in Bangladesh to assist household-level rice-processing businesses by enabling the men to buy additional paddy. Although their wives undertook additional productive processing activities, this was simply identified as an extension of reproductive work, and men still 'sold the final product and controlled the earnings' (Carr 1984). In contrast to this, in a *gari* processing project in Ghana in which the introduction of new technology enabled the women to increase their output of *gari* considerably, the women controlled the marketing. The fact that the project recognized women's productive role meant that they achieved the strategic gender need of keeping control over the resources themselves. In fact they were able to use their additional resources to buy a tractor to assist the men to put more land under cassava cultivation. In this case 'involving the women in the design of the technology undoubtedly contributed to the success of the project' (Carr 1984: 18).

Gender needs in human settlements and housing

In the planning of human settlements and housing the necessity of introducing a gender perspective is still not widely recognized, despite the fact that women, as wives and mothers, are primary users of space both in their houses and in the local community.[8]

In the planning stage, although consultation with women about housing design would ensure that their spatial needs are met, this rarely occurs. Examples of the detrimental effects of insensitive house design are widespread, particularly where 'modernization' or 'developmentalism' has resulted in radical design changes. This often affects Muslim women. Because their social life is almost entirely confined to the home, they have special needs for internal space. An example of this in Tunis, Tunisia, occurred in two low-income settlements, Mellassine, a squatter upgrading project, and Ibn Khalkoun, a planned community (Resources for Action 1982). Women were dissatisfied with house design because of the small size of the inner courtyard. Pressure on land, insensitivity to women's needs, and middle-class aspirations to European architecture had resulted in a reduction of inner courtyard area, in some cases leading to psychological depression, neuroses and even suicide among women residents.

Zoning legislation that separates residential and business activities assumes the separation of productive and reproductive roles, and is particularly problematic for women with children. Because of the need to balance roles, women are often involved in informal sector activities in or around their homes. Where zoning legislation prevents the making and selling of goods from their homes, the only solution is to do so illegally. At the Dandora Site and Service Project, in Nairobi, Kenya, restrictive zoning legislation was identified by women as the main cause of arrears in repaying building loans (Nimpuno-Parente 1987). Gender-aware changes in zoning legislation to allow household enterprises can therefore meet the practical gender need of women to earn an income.

The fact that women are the principal users of housing does not necessarily mean that they become the owners of either the house or land. Tenure is generally given to men as household heads, even where women *de facto* have primary household responsibilities. For women, tenure rights are a strategic gender need which ensure protection for themselves and their children in unstable or violent domestic situations. Without land rights women often cannot provide collateral to gain access to credit; since ownership of land represents a form of saving, women may end up without capital in the event of marital separation. Housing projects that provide for ownership regardless of the sex of the 'household head' may be difficult to design in countries where women do not legally possess rights of ownership. However, in other contexts where it is a consequence of 'tradition', there may be relatively simple means of making land-ownership rights available to women. In an upgrading project in Jordan, for example, it was found that because a woman staff member happened to be in charge of handling the title deeds in the Community Development Office, many men in the community allowed their wives to complete the relevant paperwork. As a result, title deeds ended up in the women's names, thus meeting, if unintentionally, their strategic gender need to own land. Examples of rural development schemes which have displaced women from their traditional agricultural work and their control over land are widespread, with the most common problem the failure to recognize women's productive role in the existing gender division of labour (Beneria 1979, 1982).

Gender needs in environmental planning

In their reproductive role, rural women depend on the resources of wood, water and soil for daily survival, and as such are often the primary users of the environment (Agarwal 1981, 1986). Yet their practical gender needs are often not recognized by those who utilize the environment as a productive resource. Thus Shiva, in her feminist analysis of women and the environment, describes forestry, for instance, in the following terms: 'There are in India, today, two paradigms of forestry – one life-enhancing, the other life-destroying. The life-enhancing paradigm emerges from the forest and the feminine principle; the life-destroying one from the factory and the market' (1988: 76).

The Chipko movement in the Himalayan foothills of India provides a graphic example of this conflict of needs. When the Forestry Development

Corporation employing local men sought to cut down the forests for commercial extraction, on the basis that they produce 'profit, resin and timber', local women resisted because they identified the forest as bearing 'soil, water and pure air' (Shiva 1988: 77). The solidarity between local men and women broke down; men wanted felling contracts and the women organized themselves into a co-operative to protect the forests. They provided work for women as guards to ensure that no felling went ahead (Dankelman and Davidson 1988). The community management of SIRDO waste recycling in Merida, Mexico, by contrast, was an environmental project, based on a recognition of women's triple role. In this resource conservation project designed to improve sanitation, water and waste control, women as community managers formed a committee to supervise the operation. This not only increased overall community health which concerned them as mothers and wives, but also allowed women in their productive role to sell the fertilizer produced by the project (Schmink 1984b, 1989).

Gender needs in basic services

The delivery of basic services in both rural and urban areas can have fundamental implications in terms of the gender needs met. The planning of child-care facilities provides a good example of the way in which differences in location can result in the fulfilment of different gender needs. If located at the woman's workplace it will certainly meet her practical gender need for adequate child-care facilities, essential for her to undertake waged employment. If located in the community this may encourage a sharing of responsibilities within the family, although if anyone other than the mother delivers the child it is likely to be another female member of the family. If, however, the nursery is located at the father's place of work, this provides the opportunity for meeting both practical and strategic gender needs since it involves the father in taking some of the responsibility for child care, and thereby alleviates the burden of domestic labour for the woman.

In the transport sector, one of the most critical problems faced by women is that transport services are organized to meet the needs of male workforce schedules, with buses running from the periphery to the centre during morning and evening peak periods. In a study of transport needs in Belo Horizonte, Brazil, Schmink (1982) showed that the routing of buses from the low-income settlements to the industrial areas and then into the city meant that women spent twice as much on transport costs. Their daily average travel time was three times longer than that of men. Low-income women often not only use public transport more than men but require it for multiple activities, such as school, shopping and health-related trips, in addition to work trips. Although the provision of adequate off-peak transport meets practical gender needs, it cannot meet strategic gender needs since it does not alleviate women's burden of domestic labour and child care. In addition, in many large cities the fear of male harassment prevents low-income women from using public transport, particularly late at night. Where women-only transportation is introduced, this meets the more strategic gender need of countering male violence.

In rural communities the timing of meetings can radically affect women's attendance, and consequently their capacity to gain access to important information relevant to them in both their productive and reproductive roles. Complaints by rural extension workers that women fail to attend their meetings are widespread. In a project in Botswana, for instance, meetings were held in the morning in order to ensure that male farmers, gathering in their productive and community politics role, were sober and attentive.[9] However, this timing automatically excluded women who were busy at this point of the day with essential reproductive responsibilities. Since information about family planning as well as credit schemes was intended primarily for women, rural extension workers were obliged to reschedule the meeting to an hour later in the day when women had 'free' time.

The examples cited above show the limitations of individual sectoral interventions for low-income women. Because of the necessity to balance their triple role, women require integrative strategies which cut across sectoral lines. They also reveal that the majority of planning interventions intended for women meet practical gender needs, and do not seek to change existing divisions of labour. Therefore they are not 'feminist' in content. In reality, practical gender needs remain the only specific policy target for most of those concerned with planning for women. Nevertheless, examples such as these do show that practical gender needs can be met once planners differentiate target groups not only on the basis of income, now commonly accepted, but also on the basis of gender.

The way the state has changed its policy towards women in developing countries, and the extent to which shifts in policy have occurred during the past thirty years, can best be understood through the examination of different policy approaches to women.

4   Third World policy approaches to women in development

Throughout the Third World, particularly in the past fifteen years, there has been a proliferation of policies, programmes and projects designed to assist low-income women. Until recently, however, there has been little systematic classification or categorization of these various policy initiatives, other than the informative work of Buvinic (1983, 1986). This concern for low-income women's needs has coincided historically with a recognition of their important role in development. Since the 1950s many different interventions have been formulated. These reflect changes in macro-level economic and social policy approaches to Third World development, as well as in state policy towards women. Thus the shift in policy approaches towards women, from 'welfare', to 'equity' to 'anti-poverty', as categorized by Buvinic (1983), to two other approaches which I categorize as 'efficiency' and 'empowerment' has mirrored general trends in Third World development policies, from modernization policies of accelerated growth, through basic needs strategies associated with redistribution, to the more recent compensatory measures associated with structural adjustment policies.

Wide-scale confusion still exists concerning both the definition and use of different policy approaches. Many institutions at both national government and international agency level are unclear about their policy approach to women. Often the ubiquitous, so-called 'women in development' approach has mystified rather than clarified conceptual categories. This has served to legitimize a range of approaches to women, which incorporate different underlying assumptions in relation to their practical and strategic gender needs. It is precisely because of confusions such as these that it is important to develop simple, but sufficiently rigorous, tools to enable policy-makers and planners to understand with greater clarity the implications of their interventions in terms of both their potential and limitations in assisting Third World women.

To identify the extent to which policy interventions have been appropriate to the gender needs of women, it is necessary to examine their underlying

Table 4.1 Different policy approaches to Third World women

Isssues	Welfare	Equity
Origins	Earliest approach: – residual model of social welfare under colonial administration – modernization/ accelerated growth economic development model	Original WID approach: – failure of modernization development policy – influence of Boserup and First World Feminists on Percy Amendment of UN Decade for Women
Period most popular	1950–70; but still widely used	1975–85: attempts to adopt it during the Women's Decade
Purpose	To bring women into development as better mothers: this is seen as their most important role in development	To gain equity for women in the development process: women seen as active participants in development
Needs of women met and roles recognized	To meet PGN in reproductive role, relating particularly to food aid, malnutrition and family planning	To meet SGN in terms of triple role – directly through state top-down intervention, giving political and economic autonomy by reducing inequality with men
Comment	Women seen as passive beneficiaries of development with focus on their reproductive role; non-challenging, therefore widely popular especially with government and traditional NGOs	In identifying subordinate position of women in terms of relationship to men, challenging, criticized as Western feminism, considered threatening and not popular with government

PGN = Practical gender need
SGN = Strategic gender need

Tabel 4.1 (Continued)

Anti-poverty	Efficiency	Empowerment
Second WID approach: – toned down equity because of criticism – linked to redistribution with growth and basic needs	Third and now predominant WID approach: – deterioration in the world economy – policies of economic stabilization and adjustment rely on women's economic contribution to development	Most recent approach: – arose out of failure of equity approach – Third World women's feminist writing and grassroots organization
1970s onward: still limited popularity	Post-1980s: now most popular approach	1975 onward: accelerated during 1980s, still limited popularity
To ensure poor women increase their productivity: women's poverty seen as a problem of underdevelopment, not of subordination	To ensure development is more efficient and more effective: women's economic participation seen as associated with equity	To empower women through greater self-reliance: women's subordination seen not only as problem of men but also of colonial and neo-colonial oppression
To meet PGN in productive role, to earn an income, particularly in small-scale income-generating projects	To meet PGN in context of declining social services by relying on all three roles of women and elasticity of women's time	To reach SGN in terms of triple role – indirectly through bottom-up mobilization around PGN as a means to confront oppression
Poor women isolated as separate category with tendency only to recognize productive role; reluctance of government to give limited aid to women means popularity still at small-scale NGO level	Women seen entirely in terms of delivery capacity and ability to extend working day; most popular approach both with governments and multilateral agencies	Potentially challenging with emphasis on Third World and women's self-reliance; largely unsupported by governments and agencies; avoidance of Western feminism criticism means slow, significant growth of under-financed voluntary organizations

rationale from a gender planning perspective. In this chapter different policy approaches to women in development are examined in terms of roles recognized, practical or strategic gender needs met, and the extent to which participatory planning procedures are included. (Such analysis, summarized in Table 4.1, provides the basis for the development of further principles of gender planning.)

While the policy interventions are described chronologically, from welfare through to empowerment it is recognized that the linear process that this implies is an over-simplification of reality. In practice, many of the policies have appeared more or less simultaneously. Implementing agencies have not necessarily followed any ordered logic in changing their approach, most frequently jumping from welfare to efficiency without consideration of other approaches. Similarly, different policies have particular appeal to different types of institutions. Policy-makers often favour combined policy approaches in order simultaneously to meet the needs of different constituencies. Finally, shifts in policy approach often occur not only during the formulation stage, but also during the implementation process (Buvinic 1986). Given these caveats, the following policy types described should be viewed as 'ideal types'. The purpose here is to measure how far different policies meet practical or strategic needs (see Table 4.1).

THE WELFARE APPROACH

Introduced in the 1950s and 1960s, welfare is the earliest policy approach concerned with women in developing countries. Its purpose is to bring women into development as better mothers. Women are seen as passive beneficiaries of development. The reproductive role of women is recognized and policy seeks to meet practical gender needs through that role by top-down handouts of food aid, measures against malnutrition and family planning. It is non-challenging and therefore still widely popular.

The welfare approach is the oldest and still the most popular social development policy for the Third World in general, and for women in particular. It can be identified as pre-WID. Its underlying rationale towards women reflects its origins, which are linked to the residual model of social welfare, first introduced by colonial authorities in many Third World countries prior to independence. Their concern with law and order and the maintenance of stable conditions for trade and agricultural and mineral expansion meant that social welfare was a low priority. Echoing the nineteenth-century European Poor Laws with their inherent belief that social needs should be satisfied through individual effort in the market place, administrations dealt largely with crime, delinquency, prostitution and other forms of 'deviant' behaviour.

Voluntary charity organizations in turn carried a large share of the burden of social welfare (Hardiman and Midgley 1982). Because of welfare policy's compatibility with the prevailing development paradigms of modernization, it was continued by many post-independence governments (MacPherson and Midgley 1987). On the basis that 'social welfare institutions should come into play only when the normal structure of supply, the family and the market, break down' (Wilensky and Lebeaux 1965: 138), the ministries of social welfare, created for the implementation of such residual measures for 'vulnerable' groups, were invariably weak and under-financed.

In fact it was First World welfare programmes, widely initiated in Europe after the end of World War II, specifically targeted at 'vulnerable groups', which were among the first to identify women as the main beneficiaries. As Buvinic (1986) has noted, these were the emergency relief programmes accompanying the economic assistance measures intended to ensure reconstruction. Relief aid was provided directly to low-income women, who, in their gendered roles as wives and mothers, were seen as those primarily concerned with their family's welfare. This relief distribution was undertaken by international private relief agencies, and relied on the unpaid work of middle-class women volunteers for effective and cheap implementation.

The creation of two parallel approaches to development assistance – on the one hand, financial aid for economic growth; on the other hand, relief aid for socially deprived groups – was then replicated in development policy towards Third World countries. This strategy had critical implications for Third World women. It meant that international economic aid prioritized government support for capital-intensive, industrial and agricultural production in the formal sector, for the acceleration of growth focused on increasing the productive capacity of the male labour force. Welfare provision for the family was targeted at women, who, along with the disabled and the sick, were identified as 'vulnerable' groups, and remained the responsibility of the marginalized ministries of social welfare.

In most countries these ministries and the profession of social planning, frequently seen as their mandate, were from the outset dominated by women, particularly at the lower levels. Consequently, welfare policy was, and still is, frequently identified as 'women's work', serving to reinforce social planning as soft-edged, and of lesser importance than the hard-edged areas of economic and physical planning. Further assistance was then also provided by NGOs, such as the mother's clubs created in many Third World countries, and, to a lesser extent, by bilateral aid agencies with specific mandates for women and children, such as the United Nations Children's Fund (UNICEF).

The welfare approach is based on three assumptions. First, that women are passive recipients of development, rather than participants in the development process. Secondly, that motherhood is the most important role for women in society. Thirdly, that child-rearing is the most effective role for women in all aspects of economic development. While this approach sees itself as 'family-centred' in orientation, it focuses on women entirely in terms of their reproductive role, it assumes men's role to be productive, and it identifies the mother–child dyad as the unit of concern. The main method of implementation is through 'top-down' handouts of free goods and services, and therefore it does not include women or gender-aware local organizations in participatory planning processes. When training is included it is for those skills deemed appropriate for 'non-working' housewives and mothers. In their mothering roles low-income women have been the primary targets for improving family welfare, particularly of children, through an increasing diversity of programmes, reflecting a broadening of the mandate of welfare over the past three decades.

With its origins in relief work, the first, and still the most important, concern of welfare programmes is family physical survival, through the direct provision of food aid. Generally this is provided in the short term after such natural disasters as earthquakes or famines. However, food aid has increasingly become a longer-term need for refugees seeking protection. Although the majority of refugees in camps are women, left as heads of households to care and often provide for the children and elderly, they usually do not have refugee status in their own right but only as wives within the family (Bonnerjea 1985). Projects implemented by the United Nations High Commission for Refugees (UNHCR) and NGOs most often focus on these women in their reproductive role, with special attention given to those pregnant or lactating. These are identified as a 'vulnerable' group in the same category as the elderly, orphans and the handicapped (Weeda 1987).

In the extensive international effort to combat Third World malnutrition, another emphasis of welfare programmes is nutritional education. This targets children under five years, as well as pregnant and nursing mothers. Since the 1960s, Mother–Child Health Programmes (MCH) have distributed, cooked or rationed food along with giving nutrition education at feeding centres and health clinics. In linking together additional food for children and nutrition education for mothers, MCH focuses on the mother–child dyad, and the reproductive role of women, on the assumption that extra provisions will make them better mothers. Although by the early 1980s considerable criticism had been expressed about the use of food aid to guarantee nutritional improvement of children, the focus on women in their role as mothers was not seen as problematic.[2]

Most recently, especially since the 1970s, welfare policy towards women has been extended to include population control through family planning programmes. Thus development agencies responding to the world's population 'problem' identified women, in their reproductive role, as primarily responsible for limiting the size of families. Early programmes assumed that poverty could be reduced by simply limiting fertility, to be achieved through the widespread dissemination of contraceptive knowledge and technology to women. Only the obvious failure of this approach led population planners to realize that variables relating to women's status, such as education and labour-force participation, could affect fertility differentials and consequently needed to be taken into consideration. By 1984 the World Bank's World Development Report, for instance, identified reducing infant and child mortality, educating parents (especially women) and raising rural incomes, women's employment and legal and social status, as key incentives to fertility decline (World Bank 1984). However, recognition of the links between women's autonomy over their own lives and fertility control is not widespread and women continue to be treated in an instrumental manner in population programmes. The lack of satisfactory birth-control methods, and the introduction of more invasive techniques (such as IUDs and hormonal implants) is making birth control even more 'women-centred'. As DAWN (1985) has argued, this lets men off the hook in terms of their responsibility for birth control, while increasingly placing the burden on women. Their ambivalence towards contraceptive technology will only be removed when the technology is better adapted to the social and health environments in which they are used.

Although welfare programmes for women have widened their scope considerably over the past decades, the underlying assumption is still that motherhood is the most important role for women in Third World development. This means that their major concern has been with meeting practical gender needs relating to women's reproductive role. Intrinsically, welfare programmes identify 'women' rather than lack of resources, as the problem, and place the solution to family welfare in their hands, without questioning their 'natural' role. Although the top-down handout nature of so many welfare programmes tends to create dependency rather than assisting women to become more independent, they remain popular precisely because they are politically safe, not questioning or changing the traditionally accepted role of women within the gender division of labour. Such assumptions tend to result in the exclusion of women from development programmes operated by the mainstream development agencies which provide a significant proportion of development funds (Germaine 1977). The fact that the welfare approach is not concerned to meet such strategic gender needs as the right for women to have control over their own reproduction was highlighted by a Third World women's group when they wrote:

Women know that childbearing is a social not a purely personal phenomenon: nor do we deny that world population trends are likely to exert considerable pressure on resources and institutions by the end of the century. But our bodies have become a pawn in the struggles among states, religions, male heads of households, and private corporations. Programmes that do not take the interests of women into account are unlikely to succeed.

(DAWN 1985: 42)

Although by the 1970s dissatisfaction with the welfare approach was widespread, criticism differed as to its limitations. This depended on which of the three constituencies it came from; first, in the United States, a group of mainly female professionals and researchers who were concerned with the increasing evidence that Third World development projects were negatively affecting women; second, development economists and planners who were concerned with the failure of modernization theory in the Third World; and third, the United Nations (UN), that combined both of these concerns. The voicing of these concerns led to the United Nations 1975 International Women's Year Conference. This formally 'put women on the agenda' and provided legitimacy for the proliferation of a wide diversity of Third World women's organizations, in turn leading to the UN designating 1976–85 as the Women's Decade.

During this decade the critique of the welfare approach resulted in the development of a number of alternative approaches to women: namely, equity, anti-poverty, efficiency and empowerment. The fact that these approaches share many common origins, were formulated during the same decade and are not entirely mutually exclusive, means that there has been a tendency not only to confuse them, but indeed to categorize them together as the 'women in development' (WID) approach. With hindsight, it is clear that there are significant differences between these approaches which it is important to clarify.

The lack of definition of WID has been widespread in the proliferating number of national-level WID ministries and bureaux, which implement a large number of policies under the umbrella of the WID approach (Gordan 1984). This has also been the case with bilateral and multilateral donors.[3]

THE EQUITY APPROACH

Equity is the original 'WID' approach, introduced within the 1976–85 UN Women's Decade. Its purpose is to gain equity for women in the development process. Women are seen as active participants in development. It recognizes women's triple role and seeks to meet strategic gender needs through direct state intervention, giving political and economic autonomy to women, and reducing inequality with men. It challenges women's subordinate position, has been criticized as Western feminism, is considered threatening and is unpopular with governments.

By the 1970s studies showed that although women were often the predominant contributors to the basic productivity of their communities, particularly in agriculture, their economic contribution was referred to neither in national statistics nor in the planning and implementation of development projects (Boserup 1970). At the same time new modernization projects, with innovative agricultural methods and sophisticated technologies, were negatively affecting women. These were displacing them from their traditional productive functions, and diminishing the income, status and power they had in traditional relations. Findings indicated that neo-colonialism, as much as colonialism, was contributing to the decline in women's status in developing countries.

Tinker, in her documentation of development projects that had widened the gap between men and women, argued that development planners were 'unable to deal with the fact that women must perform two roles in society whereas men perform only one' (1976; 22). She attributed the adverse impact of development on women to three types of planning error: first, errors of omission or failure to acknowledge and utilize women's productive role; second, errors that reinforced values which restrict women to the household engaged in childbearing and childrearing activities; and third, errors of inappropriate application of Western values regarding women's work (1976). On the basis of evidence such as this, the WID group in the United States challenged the prevailing assumption that modernization was equated with increasing gender equality, asserting that capitalist development models imposed on much of the Third World had exacerbated inequalities between men and women. Recognition of the damaging effects of ignoring women in USAID projects during the First Development Decade (1960–70) made the WID group work to influence USAID policy. Lobbying of Congressional hearings resulted in the 1973 Percy Amendment to the US Foreign Assistance Act, which mandated that US assistance help 'move women into their national economies' in order to improve women's status and assist the development process (Tinker 1982; Maguire 1984).

The original WID approach was in fact the equity approach.[4] This approach recognizes that women are active participants in the development process, who through both their productive and reproductive roles provide a critical, if often unacknowledged, contribution to economic growth. The approach starts with the basic assumption that economic strategies have frequently had a negative impact on women. It acknowledges that they must be 'brought into' the development process through access to employment and the market place. It therefore accepts women's practical gender need to earn a livelihood. However, the equity approach is also concerned with fundamental issues of equality which transcend the development field. As Buvinic (1986) has described, its primary concern is with inequality between men and women, in both public and private spheres of life and across socio-economic groups. It identifies the origins of women's subordination not only in the context of the family, but also in relations between men and women in the market place. Hence it places considerable emphasis on economic independence as synonymous with equity. __

In focusing particularly on reducing inequality between men and women in the gender division of labour, the equity approach meets an important strategic gender need. Equity programmes are identified as uniting notions of development and equality. The underlying logic is that women have lost ground to men in the development process. Therefore, in a process of redistribution, men have to share in a manner which entails women from all socio-economic classes gaining and men from all socio-economic classes losing (or gaining less), through positive discrimination policies if necessary. The rational consequence of this is seen to be greater equality with an accompanying increase in economic growth (Buvinic 1983). Although the approach emphasized 'top-down' legislative and other measures as the means to ensure equity, gendered consultative and participatory planning procedures were implicitly assumed. This was particularly the case since the introduction of the equity approach itself had been the consequence of the bottom-up confrontation of existing procedures by feminist women's organizations.

In fact, the theme selection for the 1975 International Women's Year (IWY) Conference showed that the equity approach, despite its identification as 'developmental', in many respects was more concerned to reflect First World feminist preoccupations with equality. Third World delegations, while acknowledging women's problems, identified development as their main concern, maintaining that this would increase women's status. Second World delegates were more concerned with peace, claiming that the capitalist system and its associated militarism was responsible for women's problems – hence the theme of Equality, Development and Peace (Stephenson 1982).

The World Plan of Action for the Implementation of the Objectives of the IWY firmly reflected the equity approach. It called for equality between men and women, required that women should be given their fair share of the benefits of development, and recognized the need for changes in the traditional role of men as well as women (UN 1976a). The Plan set the agenda for future action for the Women's Decade, with the common goal of integrating women into the development process. In reality, the interpretation of the agenda varied. This was reflected in the language used, which ranged from the definitely expressed aim to 'integrate', 'increase', 'improve' or 'upgrade' women's participation in development to the more tentatively worded desire to 'help create a more favorable climate for improving women's options in development' (World Bank 1980: 14).

Despite such rhetoric, equity programmes encountered problems from the outset. Methodologically, the lack of a single unified indicator of social status or progress of women and of baseline information about women's economic, social and political status meant that there were no standards against which 'success' could be measured (USAID 1978). Politically, the majority of development agencies were hostile to equity programmes precisely because of their intention to meet not only practical gender needs but also strategic gender needs, whose very success depended on an implicit redistribution of power. As Buvinic has commented:

> Productivity programmes for women usually require some restructuring of the cultural fabric of society, and development agencies do not like to tamper with unknown and unfamiliar social variables. As a rule of thumb they tend to believe in upholding social traditions and thus are reluctant to implement these programmes.
>
> (1983: 26)

From the perspective of the aid agency, equity programmes necessitated unacceptable interference with the country's traditions. At the same time recognition of equity as a policy principle did not guarantee its implementation in practice. In Europe the Organization for Economic Cooperation and Development Assistance Committee's (OECD/DAC) Guiding Principles to Aid Agencies for Supporting the Role of Women in Development identified 'integration' as critical for a policy on WID issues. However, in a review of European Development Assistance, Andersen and Baud (1987) argued that although policy statements of most donor countries were in general accord with the idea of equality, nevertheless at the level of policy, integration had been mainly interpreted to mean an increasing number of women in existing policies and programmes. Thus, they concluded that 'implicit in such an approach was the idea that current development models were in principle favourable to women, and that they therefore did not need to take account of women's vision or priorities' (1987: 22).

Despite their endorsement of the Plan of Action, similar antipathy was felt by many Third World governments, legitimized by their belief in the irrelevance of Western-exported feminism to Third World women. In fact, one of the outcomes of the 1975 Conference was the labelling of feminism as ethnocentric and divisive to WID. Many Third World activists felt that to take 'feminism to a woman who has no water, no food and no home is to talk nonsense' (Bunch 1980: 27), and labelled Third World socialists and feminists as bourgeois imperialist sympathizers. At the same time the fact that there was only one reference to women in the various documents of the 1970 UN New International Economic Order Conference revealed that the importance of women was still identified in terms of their biological role by those formulating policies for the Third World.

In a climate of widespread antagonism to many of its underlying principles from development agencies and Third World governments alike, the equity approach has been effectively dropped by the majority of implementing agencies. However, its official endorsement in 1975 ensured that it continues to provide an important framework for those working within government to improve the status of women through official legislation, on issues such as those described in Chapter 3. Tinker and Jaquette (1987), in reviewing the 1976–85 Women's Decade conference documents, noted that the goal of legal equality of women had been accepted as a minimum basis of consensus from which to begin the discussion of more controversial issues. This included the rights of divorce, of custody of children, property, credit, voting and other citizen rights.

Significant though the ratification of such legislation is, it is necessary to recognize that it meets potential strategic gender needs, rather than actual needs. As illustrated in Chapter 3, property rights, arranged marriages, dowry and child custody rights provide much cited examples of the highly sensitive strategic gender needs which are often still curtailed by custom, even when amended by law. Even the incorporation of practical gender needs into the mainstream of development plans does not guarantee their implementation in practice. Mazumdar (1979) noted that the incorporation of women's concerns into the framework of India's Six Year Plan indicated India's constitutional commitment to equality of opportunity. Such constitutional inclusions, however, in no way ensured practical changes. In her opinion these are largely a function of the strength of the political power base of organized women's groups. Ultimately, the equity approach has been constructed to meet strategic gender needs through top-down legislative measures. The bottom-up mobilization of women into political pressure groups to ensure that policy becomes action is the mandate of the empowerment approach, developed by Third World women, and described later.

THE ANTI-POVERTY APPROACH

Anti-Poverty is the second WID approach, the 'toned down' version of equity, introduced from the 1970s onwards. Its purpose is to ensure that poor women increase their productivity. Women's poverty is seen as the problem of underdevelopment, not of subordination. It recognizes the productive role of women, and seeks to meet practical gender needs to earn an income, particularly through small-scale income-generating projects. It is most popular with NGOs.

The anti-poverty approach to women can be identified as the second WID approach, in which economic inequality between women and men is linked not to subordination but to poverty. The emphasis thus shifts from reducing inequality between men and women, to reducing income inequality. Women's issues are separated from equity issues and linked instead to the particular concern for the majority of Third World women, as the 'poorest of the poor'. Buvinic (1983) has argued that this is a toned-down version of the equity approach, arising out of the reluctance of development agencies to interfere with the manner in which relations between men and women are constructed in a given society. However, this shift also coincided with the end of the unsuccessful First Development Decade, and the formulation of alternative models of Third World economic and social development.

By the early 1970s it was widely recognized that modernization theory, with its accelerated growth strategies based on maximizing GNP, had failed, either to redistribute income or to solve the problems of Third World poverty and unemployment. Contrary to predictions about the positive welfare effects of rapid economic growth, financial benefits had not 'trickled down' to the poor. An early initiative was the International Labour Organization's (ILO) World Employment Programme in which employment became a major policy objective in its own right. The 'working poor' were identified as the target group requiring particular attention, and the informal sector with its assumed autonomous capacity to generate employment was seen as the solution (Moser 1978, 1984). In 1972 the World Bank officially shifted from a preoccupation with economic growth to a broader concern with the eradication of absolute poverty and the promotion of 'redistribution with growth'. Integral to this was the 'basic needs strategy', with its primary purpose to meet 'basic needs' such as food, clothing, shelter and fuel, as well as social needs such as education, human rights and 'participation' in social life through employment and political involvement (Ghai 1978; Streeton *et al.* 1981). Low-income women were identified as one particular 'target group' to be assisted in escaping absolute deprivation: first, because the failure of 'trickle-down' was partially attributed to the fact that women had been ignored in previous development plans; and secondly, because of the traditional importance of women in meeting many of the basic needs of the family (Buvinic 1982).

The anti-poverty policy approach to women focuses mainly on their productive role, on the basis that poverty alleviation and the promotion of balanced economic growth requires the increased productivity of women in low-income households. Underlying this approach is the assumption that the origins of women's poverty and inequality with men are attributable to their lack of access to private ownership of land and capital, and to sexual discrimination in the labour market. Consequently, it aims to increase the employment and income-generating options of low-income women through better access to productive resources. The preoccupation of basic needs strategies with population control also resulted in increasing recognition that education and employment programmes could simultaneously increase women's economic contribution and reduce fertility.[5]

One of the principal criticisms of employment programmes for women is that since they have the potential to modify the gender division of labour within the household, they may also imply changes in the balance of power between men and women within the family. In anti-poverty programmes this redistribution of power is said to be reduced, because the focus is specifically on low-income women, and because of the tendency to encourage projects in sex-specific occupations in which women are concentrated, or to target only women who head households. The fear, however, that programmes for low-income women may reduce the already insufficient amount of aid allocated to low-income groups in general means that Third World governments have remained reluctant to allocate resources from national budgets to women. Frequently, the preference is to allocate resources at the family or household level, despite the fact that they generally remain in the hands of the male head of household.

While income-generating projects for low-income women have proliferated since the 1970s, they have tended to remain small in scale, to be developed by NGOs (most frequently all-women in composition), and to be assisted by grants, rather than loans, from international and bilateral agencies. Most frequently they aim to increase productivity in activities traditionally undertaken by women, rather than to introduce women to new areas of work, with a preference for supporting rural-based production projects as opposed to those in the service and distribution sectors, which are far more widespread in the urban areas of many developing countries.[6]

Considerable variation has been experienced in the capacity of such projects to assist low-income women to generate income. Buvinic (1986) has highlighted the problems experienced by anti-poverty projects in the implementation process, due to the preference to shift towards welfare-orientated projects. However, such projects also experience considerable constraints in the formulation stage. In theory, 'basic needs' assumed a participatory approach, yet in practice anti-poverty projects for women rarely included participatory planning procedures; mechanisms to ensure that women and gender-aware organizations be included remained undeveloped. In the design of projects, fundamental conditions to ensure viability are often ignored, including access to easily available raw materials, guaranteed markets and small-scale production capacity (Schmitz 1979; Moser 1984). Despite widespread recognition of the limitations of the informal sector to generate employment and growth in an independent or evolutionary manner, income-generating projects for women continue to be designed as though small-scale enterprises have the capacity for autonomous growth (Schmitz 1982; Moser 1984).

In addition, the particular constraints that women experience in their gendered roles are also frequently ignored. These may include problems of perception in separating reproductive from productive work, as well as those associated with 'balancing' productive work alongside domestic and child-care responsibilities. In many contexts there are cultural constraints that restrict women's ability to move freely outside the domestic arena and therefore to compete equally with men running similar enterprises (Moser 1981). Where men control household financial resources, women are unable to save unless special safe facilities are provided (Sebsted 1982). Equally, where women cannot obtain equal access to credit, such as through lack of collateral, they are often unable to expand their enterprises unless non-traditional forms of credit are available to them (Bruce 1980; IWTC 1985). Finally, the tendency to distinguish between micro-enterprise projects for men, and income-generating projects for women, is indicative of the prevailing attitude, even among many NGOs, that women's productive work is of less importance than men's, and is undertaken as a secondary earner or 'for pocket money'.

Anti-poverty income-generating projects may provide employment for women, and thereby meet practical gender needs, by augmenting their income, but unless employment leads to greater autonomy it does not meet strategic gender needs. This is the essential difference between the equity and anti-poverty approaches. In addition, the predominant focus on the productive role of women in the anti-poverty approach means that their reproductive role is often ignored. Income-generating projects which assume that women have 'free time' often only succeed by extending their working day and increasing their triple burden. Unless an income-generating project also alleviates the burden of women's domestic labour and child care – for instance, through the provision of adequate socialized child caring – it may fail even to meet practical gender need to earn an income.

THE EFFICIENCY APPROACH

Efficiency is the third, and now predominant WID approach, particularly since the 1980s debt crisis. Its purpose is to ensure that development is more efficient and effective through women's economic contribution. Women's participation is equated with equity for women. It seeks to meet practical gender needs while relying on all of women's three roles and an elastic concept of women's time. Women are seen primarily in terms of their capacity to compensate for declining social services by extending their working day. It is very popular as an approach.

Although the shift from equity to anti-poverty has been well documented, the identification of WID as efficiency has passed almost unnoticed. Yet, I would argue the efficiency approach is now the predominant approach for those working within a WID framework – indeed, for many it may always have been. In it the emphasis has shifted away from women and towards development, on the assumption that increased economic participation for Third World women is automatically linked with increased equity. This has allowed organizations such as USAID, the World Bank and OECD to propose that an increase in women's economic participation in development links efficiency and equity together. Amongst others, Maguire (1984) has argued that the shift from equity to efficiency reflected a specific economic recognition of the fact that 50 percent of the human resources available for development were being wasted or under-utilized. Although the so-called development industry realized that women were essential to the success of the total development effort, it did not necessarily follow that development improved conditions for women.[7] The assumption that economic participation increases women's status and is associated with equity has been widely criticized. Problems such as lack of education and under-productive technologies have also been identified as the predominant constraints affecting women's participation.

The shift towards efficiency coincided with a marked deterioration in the world economy, occurring from the mid-1970s onwards, particularly in Latin America and Africa, where the problems of recession were compounded by falling export prices, protectionism and the mounting burden of debt. To alleviate the situation, economic stabilization and adjustment policies designed by the International Monetary Fund (IMF) and the World Bank have been implemented by an increasing number of national governments. These policies, through both demand management and supply expansion, lead to the reallocation of resources to enable the restoration of a balance of payments equilibrium, an increase in exports and a rejuvenation in growth rates.

With increased efficiency and productivity as two of the main objectives of Structural Adjustment Policies (SAPs), it is no coincidence that efficiency is the policy approach towards women which is currently gaining popularity amongst international aid agencies and national governments alike. Again top-down in approach, without gendered participatory planning procedures, in reality SAPs often simply mean a shifting of costs from the paid to the unpaid economy, particularly through the use of women's unpaid time. While the emphasis is on women's increased economic participation, this has implications for women not only as reproducers, but also increasingly as community managers. In the housing sector, for instance, one such example is provided by 'site and service' and upgrading projects with self-help components which now regularly include women in the implementation phase. This is a consequence of the need for greater efficiency: not only are women as mothers more reliable than men in repaying building loans, but also as workers they are equally capable of self-building alongside men, while as community managers they have shown far greater commitment than men in ensuring that services are maintained (Fernando 1987; Nimpuno-Parente 1987).

Disinvestments in human resources, made in the name of greater efficiency in IMF and World Bank 'conditionality' policies, have resulted in declines in income levels, severe cuts in government social expenditure programmes, particularly health and education, and reductions in food subsidies. These cuts in many of the practical gender needs of women are seen to be cushioned by the elasticity of women's labour in increasing self-production of food, and changes in purchasing habits and consumption patterns. In fact, underlying many SAPs, as Elson has identified, are three 'kinds of male bias' (Elson 1991: 6; Moser 1992a). The first male bias, as described above, focuses on the unpaid domestic work necessary for reproducing and maintaining human resources. It concerns the extent to which SAPs implicitly assume that processes carried out by women in such unpaid activities as caring for children, gathering fuel, processing food, preparing meals and nursing the sick will continue regardless of the way in which resources are reallocated. For SAPs define economies only in terms of marketed goods and services and subsistence cash production and exclude women's reproductive work. This raises the question as to how far SAPs are only successful at the cost of longer and harder working days for women, who are forced to increase their labour both within the market and the household. Preoccupation has been expressed regarding the extent to which women's labour is infinitely elastic, or whether a breaking point may be reached when their capacity to reproduce and maintain human resources may collapse (Jolly 1987).

Moreover, the issue not only concerns the elasticity of time, but also the balancing of time. Evidence from a longitudinal study of a low-income community in Guayaquil, Ecuador, showed that the real problem was not the length of time women worked, but the way, under conditions of recession and adjustment, they were forced to change the balance of their time between activities undertaken in each of their triple roles. Over the past decade these low-income women have always worked between twelve and eighteen hours per day, depending on such factors as the composition of the household, the time of year and their skills. Therefore, the hours worked have not changed fundamentally. What has changed is the time allocated to different activities. The need to gain access to resources has forced women to allocate increasing time to productive and community managing activities, at the expense of reproductive activities, which in many cases have become a secondary priority delegated wherever possible to daughters or other female household members. The fact that paid work and unpaid work are competing for women's time has important impacts on children, on women themselves and on the disintegration of the household (Moser 1992a).

The problem of balancing time is also of importance in relation to a second 'male bias' in SAPs. This involves ignoring barriers to labour re-allocation in policies designed to switch from non-tradables to tradables, by offering incentives to encourage labour-intensive manufacturing, and, particularly in sub-Saharan Africa, crops for export. In the urban sector, gender barriers to the re-allocation of labour have often meant greater unemployment for men displaced from non-tradables, while for any women drawn into export-orientated manufacturing, they have meant extra work, as factory employment is added to the unpaid domestic work which unemployed men remain reluctant to undertake. In rural areas the introduction of export-orientated crops has often meant increased agricultural work for women with less time for the production of subsistence family crops, resulting in both increased intra-household conflict and in worrying consequences for children's nutrition levels (Evans and Young 1988; Feldman 1989) (see Chapter 2, section beginning on page 18).

The third 'male bias' concerns the household as the social institution which is the source of the supply of labour. This concerns the assumption of an equal intra-household distribution of resources, which in turn means that changes in resource allocations in income, food prices and public expenditure, accompanying stabilization and SAPs, affect all members of the household in the same way (Elson 1991). Here policy assumes that the household has a 'joint utility' or 'unified family welfare' function with a concern to maximize the welfare of all its members, even if it assumes the altruism of benevolent dictatorship. Consequently, planners have treated it as 'an individual with a single set of objectives' (Evans 1989; Elson 1992) (see also Chapter 2, section beginning on page 18).

Until recently, structural adjustment has been seen purely as an economic issue, and evaluated in economic terms (Jolly 1987). Although documentation regarding its social costs is still unsystematic, it does reveal a serious deterioration in living conditions of low-income populations resulting from a decline in income levels. A gender-differentiated impact on intra-household resource distribution, with particularly detrimental effects on the lives of children and women, is also apparent (Cornia *et al.* 1987, 1988;

Afshar and Dennis 1992). Within the household a decline in consumption often affects women more than men. The introduction of charges for education and health care can reduce access more severely for girls than for boys. The capacity of the household to shoulder the burden of adjustment can have detrimental effects in terms of human relationships, expressed in increased domestic violence, mental health disorders and increasing numbers of women-headed households resulting from the breakdown in nuclear family structures (UNICEF n.d.).

UNICEF's widely publicized plea to devise adjustment policies 'with a human face' now challenges the efficiency basis of IMF and World Bank policy. It argues that women's concerns, both in the household and in the workplace, need consciously to be made part of the formulation of adjustment policies. This in turn will require the direct involvement of women in both the definition of development and the adjustments in its management (Jolly 1987). On paper, UNICEF's current recommendations to assist low-income women would appear highly laudable. Yet optimism that an international agency has the capacity to effect policy measures designed to increase the independence of women must be treated with caution.

This point can be illustrated through the appraisal of some of the recent compensatory policies endorsed by UNICEF. These are designed to protect basic health and nutrition of the low-income population during adjustment, before growth resumption enables them to meet their basic needs independently. In a number of nutrition interventions, such as targeted food subsidies and direct feeding for the most vulnerable, it is assumed that women in their community managing role will take responsibility for the efficient delivery of such services. For example, in Lima, Peru, the Vaso de Leche (Glass of Milk) direct feeding programme, which provides a free glass of milk to young children in the low-income areas of the city, is managed by women in their unpaid time. Similarly, the much-acclaimed communal kitchen organizations which receive targeted food subsidies depend on the organizational and cooking ability of women to ensure that the cooked food reaches families in the community (Sara-Lafosse 1984). While both programmes are aimed at improving the nutritional status of the population, especially the low-income groups, this is achieved through reliance on women's unpaid time (Cornia *et al.* 1988).

These examples illustrate the fact that the efficiency approach relies heavily on the elasticity of women's labour in both their reproductive and community managing roles. It only meets practical gender needs at the cost of longer working hours and increased unpaid work. In most cases this approach fails to reach any strategic gender needs. Because of the reductions in resource allocations, it also results in a serious reduction in the practical gender needs met.

THE EMPOWERMENT APPROACH

Empowerment is the most recent approach, articulated by Third World women. Its purpose is to empower women through greater self-reliance. Women's subordination is seen not only as the problem of men but also of colonial and neo-colonial oppression. It recognizes women's triple role, and seeks to meet strategic gender needs indirectly through bottom-up mobilization around practical gender needs. It is potentially challenging, although it avoids the criticism of being Western-inspired feminism. It is unpopular except with Third World women's NGOs and their supporters.

The fifth policy approach to women is that of empowerment. It is still neither widely recognized as an 'approach' nor documented as such, although its origins are by no means recent. Superficially it may appear synonymous with the equity approach, with references often made to a combined equity/ empowerment approach. In many respects empowerment developed out of dissatisfaction with the original WID as equity approach, because of its perceived co-option into the anti-poverty and efficiency approaches. However, the empowerment approach differs from the equity approach. This relates not only in its origins, but also in the causes, dynamics and structures of women's oppression which it identifies, and in terms of the strategies it proposes to change the position of Third World women.

The origins of the empowerment approach are derived less from the research of First World women, and more from the emergent feminist writings and grassroots organizational experience of Third World women; it accedes that feminism is not simply a recent Western urban middle-class import. As Jayawardena (1986) has written, the women's movement was not imposed on women by the United Nations or Western feminists, but has an independent history. The empowerment approach acknowledges inequalities between men and women, and the origins of women's subordination in the family. But it also emphasizes the fact that women experience oppression differently according to their race, class, colonial history and current position in the international economic order. It therefore maintains that women have to challenge oppressive structures and situations simultaneously at different levels.

The empowerment approach questions some of the fundamental assumptions concerning the interrelationship between power and development that underlie previous approaches. It acknowledges the importance for women to increase their power. However, it seeks to identify power less in terms of domination over others (with its implicit assumption that a gain for women implies a loss for men), and more in terms of the capacity of women to increase their own self-reliance and internal strength. This is identified as the right to determine choices in life and to influence the direction of change, through the ability to gain control over crucial material and non-material resources. It places far less emphasis than the equity approach on increasing women's 'status' relative to men. It thus seeks to empower women through the redistribution of power within, as well as between, societies. It also questions two underlying assumptions in the equity approach: first, that development necessarily helps all men; and secondly, that women want to be 'integrated' into the mainstream of Western designed development, in which they have no choice in defining the kind of society they want (UNAPCWD 1979).

The best-known articulation of the empowerment approach has been made by the Development Alternatives with Women for a New Era (DAWN). This is a loose formation of individual women and women's groups set up prior to the 1985 World Conference of Women in Nairobi.[8] Their purpose has been not only to analyze the conditions of the world's women, but also to formulate a vision of an alternative future society, which they identify as follows:

> We want a world where inequality based on class, gender and race is absent from every country and from the relationships among countries. We want a world where basic needs become basic rights and where poverty and all forms of violence are eliminated. Each person will have the opportunity to develop her or his full potential and creativity, and women's values of nurturance and solidarity will characterize human relationships. In such a world women's reproductive role will be re-defined: childcare will be shared by men, women and society as a whole ... only by sharpening the links between equality, development and peace, can we show that the 'basic rights' of the poor and the transformations of the institutions that subordinate women are inextricably linked. They can be achieved together through the self-empowerment of women.
>
> (1985: 73–5)[9]

Using time as a basic parameter for change, DAWN distinguishes between long-term and short-term strategies. Long-term strategies are needed to break down the structures of inequality between genders, classes and nations. Fundamental requisites for this process include national liberation from colonial and neo-colonial domination, shifts from export-led strategies in agriculture and greater control over the activities of multinationals. Short-term strategies are identified as necessary to provide ways of responding to current crises. Measures to assist women include food production through the promotion of a diversified agricultural base, as well as in formal and informal sector employment.

Although short-term strategies correspond to practical gender needs, long-term strategies contain a far wider agenda than do strategic gender needs, with national liberation identified as a fundamental requisite for addressing them. DAWN in their description of this approach, however, do not identify the means to ensure that once national liberation has been achieved, women's liberation will follow. As discussed in Chapter 3, recent liberation and socialist struggles in countries such as Cuba, Nicaragua and Zimbabwe have shown this not necessarily to have been the case (Murray 1979a, 1979b; Molyneux 1981, 1985b). One of the reasons why the categorization of practical and strategic gender needs does not consider time as a determinant of change lies in the implicit, underlying assumptions that short-term change leads to long-term transformation. In the same way it cannot be assumed that meeting practical strategic gender needs will automatically result in the satisfaction of strategic gender needs.

The new era envisaged by DAWN also requires the transformation of the structures of subordination that have been so inimical to women. Changes in law, civil codes, systems of property rights, control over women's bodies, labour codes and the social and legal institutions that underwrite male control and privilege are essential if women are to attain justice in society. These strategic gender needs are similar to those identified by the equity approach. It is in the means of achieving such needs that the empowerment approach differs most fundamentally from previous approaches. Recognition of the limitations of top-down government legislation actually, rather than potentially, to meet strategic gender needs has led adherents of the empowerment approach to acknowledge that their strategies will not be implemented without the sustained and systematic efforts of women's organizations and like-minded groups. Hence it explicitly includes gendered consultative and participatory planning procedures. Important entry points for leverage identified by such organizations are therefore not only legal changes but also political mobilization, consciousness raising and popular education. All of these are mechanisms to ensure that women and gender-aware organizations are included in the planning process.

In its emphasis on women's organizations, the empowerment approach might appear similar to the welfare approach, which also stressed the importance of women's organizations. This has led some policy-makers to conflate the two approaches. However, the welfare approach recognizes only the reproductive role of women and utilizes women's organizations as a top-down means of delivering services. In contrast, the empowerment approach recognizes the triple role of women and seeks through bottom-up women's organizations to raise women's consciousness to challenge their subordination. In fact, Third World women's organizations form a continuum. This ranges from direct political action, through exchanging research and information, to the traditional service-orientated organizations with their class biases and limited scope for participatory action. While acknowledging the valuable function of different types of organizations, the empowerment approach seeks to assist the more traditional organizations to move towards a greater awareness of feminist issues. Thus Sen (1990) acknowledges that the perceptions of the individual interests of women tend to be merged with the notion of family well-being. The 'political agency' of women may be sharpened by their greater involvement with the outside world.

Another important distinction between the empowerment and equity approaches is the manner in which the former seeks to reach strategic gender needs indirectly through practical gender needs. The very limited success of the equity approach to confront directly the nature of women's subordination through legislative changes has led the empowerment approach to avoid direct confrontation. It utilizes practical gender needs as the basis on which to build a secure support base, and a means through which strategic needs may be reached. The following examples of Third World women's organizations are much quoted 'classics' of their kind, which have provided important examples for other groups of the ways in which practical gender needs can be utilized as a means of reaching strategic gender needs.

In the Philippines, GABRIELA (an alliance of local and national women's organizations) ran a project which combined women's traditional task of sewing tapestry with a non-traditional activity, the discussion of women's legal rights and the constitution. A nation-wide educational 'tapestry-making drive' enabled the discussion of rights in communities, factories and schools, with the end product a 'Tapestry of Women's Rights' seen to be a liberating instrument (Gomez 1986).

A feminist group in Bombay, India, the 'Forum against Oppression of Women' first started campaigning in 1979 on such issues as rape and bride-burning. However, with 55 percent of the low-income population living in squatter settlements, the Forum soon realized that housing was a much greater priority for local women, and, consequently, soon shifted its focus to this issue. In a context where women by tradition had no access to housing in their own right, homelessness, through breakdown of marriage or domestic violence, was an acute problem, and the provision of women's hostels a critical practical gender need. Moreover, mobilization around homelessness also raised consciousness of the patriarchal bias in inheritance legislation as well as in the interpretation of housing rights. In seeking to broaden the problem from a 'women's concern' and to raise men's awareness, the Forum has become part of a nation-wide alliance of NGOs, lobbying national government for a National Housing Charter. Through this alliance the Forum has ensured that women's strategic gender needs relating to housing rights have been placed on the mainstream political agenda, and have not remained simply the concern of women.

Conflicts often occur when empowered women's organizations succeed in challenging their subordination. One widely cited example is the Self-Employed Women's Organization (SEWA) started in Ahmedabad, India, in 1972 by a group of self-employed women labourers. It initially struggled for higher wages and for the defence of members against police harassment and exploitation by middlemen. At first, with the assistance of the male-dominated Textile Labour Association (TLA), SEWA established a bank, as well as providing support for low-income women such as skill training pro-grammes, social security systems, production and marketing co-operatives (Sebsted 1982). It has been said that the TLA expelled SEWA from its organization, not only because the TLA leaders felt increasingly threatened by the women's advance towards self-independence, but also because their methods of struggle, in opposition to TLA policy of compromise and col-laboration, provided a dangerous model for male workers (Karl 1983). SEWA has survived considerable setbacks in its development largely due to its widespread membership support. The fact that it has developed into a movement has made it increasingly difficult to eliminate. In addition, at various times the grant support SEWA has received from international agencies has assisted in giving the organization a level of independence within the local political context.

As highlighted by DAWN, 'empowering ourselves through organization' has been a slow global process, accelerating during and since the Women's Decade. A diverse range of women's organizations, movements, networks and alliances have developed. These cover a multitude of issues and pur-poses. Common interests range from disarmament at the international level to mobilization around specific laws and codes at the national level. All share a similar commitment to empower women, and a concern to reject rigid bureaucratic structures in favour of non-hierarchical open structures, al-though they are not necessarily the most efficient organizational form. Experience to date has shown that the most effective organizations have been those that started around concrete practical gender needs relating to health, employment and basic service provision, but which have been able to utilize concerns such as these as a means to reach specific strategic gender needs. In Chapter 9 this issue is further examined with the categorization of the range of women's organizations.

The potentially challenging nature of the empowerment approach has meant that it remains largely unsupported either by national governments or bilateral aid agencies. Despite the widespread growth of Third World groups and organizations, whose approach to women is essentially one of empower-ment, they remain under-funded, reliant on the use of voluntary and unpaid women's time, and dependent on the resources of those few international

NGOs and First World governments prepared to support this approach to women and development.[10]

It is clear that the 'room for manoeuvre' still remains limited, with welfare, and more recently efficiency, the predominant policy approaches endorsed by most governments and international agencies. With increasing political and ideological control in many contexts, severe difficulties continue to be encountered in shifting policy towards the anti-poverty, equity or empowerment approach. However, there are also individuals and groups involved in changing policy approaches; government and aid agency personnel who argue that a 'gendered' efficiency approach can also be the means, with a hidden agenda, to empower women; the proliferating number of under-financed, small-scale Third World women's organizations in which women are increasingly struggling not only to meet practical gender needs but also to raise consciousness to struggle for strategic gender needs.

Part One of this book has provided the conceptual rationale for gender planning. This is based on the identification of the triple role of women, the fundamental analytical distinction between practical and strategic gender needs, and the identification of five different policy approaches to WID, which differ in terms of roles recognized and gender needs met. In Chapter 5, the methodological tools deriving from these principles are further elaborated in a description of gender planning. This is a new planning tradition, which in incorporating gender into planning, challenges current planning stereotypes.

Part II

The gender planning process and the implementation of planning practice

5 Towards gender planning
A new planning tradition and planning methodology

BACKGROUND: A BRIEF OUTLINE OF PLANNING TRADITIONS AND METHODOLOGIES

Part One of this book identified the ways in which current assumptions about women and men in society result in the formulation and implementation of policies, programmes and projects that ignore, disadvantage or discriminate against Third World women. Planners are unable to deal with the 'whole' economy – that is, with both market and non-market relations – and with gender divisions of labour. It is these that provide the conceptual rationale for the identification of gender planning as a planning tradition in its own right. This chapter describes the emerging tradition of gender planning, and outlines its methodological procedures, tools and techniques. Its purpose is to propose a new planning framework that can effectively aid the goal of the emancipation of women, through strategies to challenge and overcome oppressive roles and relationships.

In order to discuss gender planning within the broader perspective of Third World planning, it is useful to clarify several commonly confused issues. First, a distinction should be made between a planning tradition and a planning methodology. A planning tradition is a particular form of planning, with its own focus and objectives, knowledge base, agenda, process and organization. In contrast, a planning methodology is the process of providing organized technical guidance for such action (Safier 1990). Secondly, and following from this, is the need to acknowledge that over time different planning traditions have used different planning methodologies. Thirdly, it is necessary to recognize that planning methodologies differ concerning the extent to which they identify planning as a set of technical or political procedures.

Since planning was first identified as a professional activity, a range of different traditions, each with an associated methodology and relative perception relating to the 'neutrality' of the activity, has evolved. Safier (1990) categorizes these planning traditions into three broad 'generations' or groups, and distinguishes them as follows. First, there are the 'classical' traditions, concerned with physical and spatial problems of city growth. These started around the 1890s with emphases on urban design, town planning and land-use planning, followed, from the 1930s onwards, by regional planning and transport planning. The planning methodology most widely identified with these traditions was the traditional survey–analysis–plan, or so-called 'blueprint' approach to planning (see Table 5.1). Product-orientated in its focus on plans, its best-known form was the national plan, adopted in many Third World countries. The methodology comprises straightforward stages from survey to analysis, both of which social scientists undertake. Implementation of the plan follows. Most frequently spatial in nature, engineers and architects usually execute it. The methodology assumes a consensus on values and policy directions in the management of change, encapsulated in the notion of 'public interest' (Healey 1989).

Table 5.1   Planning methodology 1: the blueprint plan

Survey   →   Analysis   →   Plan

In the past decade, planning theorists and practitioners have rejected the 'blueprint' approach. This is both because of the *naïveté* of its social and economic ideas, and the political and intellectual assumptions upon which it is based. Healey (1989) has argued that the problem with blueprint planning has been its effective assumption that state bureaucrats, in charge of the development process from the planning to implementation stages, could translate knowledge about economic, social and environmental needs into spatial and physical forms. The assumption was that without making complex value choices on the way, such knowledge could be transformed into policies, plans, programmes and projects. This led Healey to conclude that the blueprint approach was both politically authoritarian and epistemologically naïve.

Second come the 'applied' traditions, which developed during the 1950s and 1960s with the increasing complexities of the global economic system. Their concern shifted from the spatial and physical domain to the underlying economic, social and governmental systems that generated contemporary patterns of growth. They subsequently resulted in social and economic planning traditions, at both project and corporate levels. With their primary concern the promotion of rationality in planning, applied traditions base themselves on procedural planning theory. These derive from a general system model that ascribes to planning certain societal tasks to be pursued through a problem-solving technology (see Healey *et al.* 1982).

The applied traditions characterized planning as a set of rational procedures and methods for decision-making. The so-called rational comprehensive methodology of procedural planning consists of several logical stages. These start with problem definition, and develop through data collection and processing. The formulation of goals and objectives and the design of alternative plans follows. Finally, there are the processes of decision-making, implementation, monitoring and feedback. Project planning is one such applied tradition to adopt rational comprehensive planning methodology. This emerged as a pragmatic alternative to the failure of more traditional planning traditions. Donor agencies, in particular, have widely adopted it. Although terminology varies, with procedures extended or amended, nevertheless the generalized project cycle, identified by Baum, is still widely followed:

> Each project passes through a cycle that with some variations, is common to all . . . each phase leads to the next, and the last phases in turn produce new project approaches and ideas and lead to the identification of new projects, making the cycle self-renewing.
>
> (Baum 1982: 5)

Table 5.2 illustrates the different stages of the rational comprehensive methodology of procedural planning, with the project-cycle terminology included in brackets.

Table 5.2 Planning methodology 2: rational comprehensive planning with particular reference to the project cycle

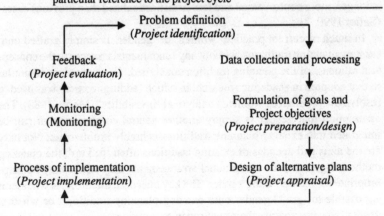

Both procedural planning and its application in practice have been widely criticized. Thomas (1979), for example, has commented that it focuses on the means of planning rather than on the end product. The assumption is that a distinctive type of planning thought and action can exist, without reference to any particular object. Its adoption within any societal context, he argues, is both 'contentless' and 'contextless'. It is 'contentless' in that it specifies cognitive and behavioural procedures but does not investigate their content. As Hambleton (1986) maintains, this depoliticizes planning. It is 'context-less' in that it fails to locate planning in its historical context. Moreover, the separation of the *content* of planning from the *context* in which it is located results in the separation of facts from values; rational comprehensive planning places prescriptive emphasis on organizational structures and methods to promote rational decision-making. It has also been criticized for its assumption that what is rational can be proved scientifically or logically. Criticisms such as these have resulted in the elaboration of several other planning methodologies such as incremental, advocacy, action and strategic planning. Nevertheless, rational comprehensive planning, in a modified form, remains the predominant planning methodology. Institutional structures as diverse as governments, donor agencies and NGOs all use it for their planning procedures (Healey *et al.* 1982).

Third come the 'transformative' traditions, still currently undergoing evolution and yet to be fully established, such as development, cultural and environmental planning, to which may be added gender planning. Equally underdeveloped are the planning methodologies associated with these traditions, which require far more 'transformative' procedures than is the case in existing methodologies. Thus, Safier has argued that 'These traditions are marked by their potential "transformative" impact on the way we perceive and wish to experience life in all its aspects. Consequently their emergence is longer, more controversial and less firmly articulated than previous cases' (Safier 1990: 7).

In much of current practice 'women' or 'gender' is simply grafted onto existing planning traditions, without any fundamental changes to the conceptual rationale of the planning tradition concerned. This procedure is similar to that adopted in academic research, in which 'adding women' was used to resolve the problem of women's analytical 'invisibility' (Moore 1988). The assumption that gender is simply another neutral component that can be integrated into existing planning traditions is highly problematic. Not only are the aims and agendas of existing traditions often 'full up'; the planning methodologies and the institutional arrangements they utilize may be inappropriate for incorporating gender. The key question, therefore, is whether it is possible to 'graft' gender onto existing planning traditions, or whether gender concerns require their own separate tradition.

Just as feminists found 'that the goals of the women's movement could not be fulfilled by the "add-women-and-stir method"' (Boxer 1982), so many planners recognize the need to develop gender planning as a planning tradition in its own right. The general goal is the emancipation of women and their release from subordination, with the aim of achieving gender equity, equality and empowerment through meeting practical and strategic needs. However, the force to create such a tradition does not derive simply from a dissatisfaction with current technical procedures in planning. Above all, it is a response to the powerful social and political movement for emancipation generated by women themselves. The purpose of gender planning then is to provide the means by which to operationalize this political concern, and to ensure that it becomes institutionalized in planning practice.

ISSUES IN GENDER PLANNING

Gender planning includes several critical characteristics. First, it is both political and technical in nature. Secondly, it assumes conflict in the planning process. Thirdly, it involves transformative processes. Fourthly, it characterizes planning as 'debate'.[1] Each of these characteristics is examined below.

Analysis of the underlying reasons for current planning assumptions relates to the roles and needs of women and men in society. As outlined in Part One, this highlights the overtly political nature of gender planning. It challenges outright the view that a gender planning methodology can simply adopt an existing 'neutral' and universally applicable set of technical procedures. It also raises the question as to the extent to which planners themselves play a role in shaping people's demands. The separation of values and facts allows for the precise distinction between client and expert. However, in reality planners are experts providing 'value-laden' advice, and, along with other actors, cannot be neutral (Healey 1989). Furthermore, given that the goal of gender planning is the emancipation of women, the political nature of the planning tradition is explicit. The rationale for the planning process then requires 'negotiation' as the basis for its agenda. Rational comprehensive planning techniques, appropriate for planning traditions based on consensus, are, therefore, inadequate as the methodology for gender planning. It requires a methodology in which the political dimension of negotiation is its central concern.

In identifying such a methodology, two recently articulated methodologies can contribute to gender planning. One of these is the *political economy* approach to planning, which focuses on the redistribution of resources through structural change. Based on a conflict model of society, it argues that the unequal distribution of resources is the inevitable outcome of fundamental class conflict. The role of the state is often one of an integrating mechanism, designed to preserve its own interests. Thus, the state is not neutral but a major actor in the conflict inherent in the planning process. In Chapter 3, the most important principle of gender planning identified was the differentiation between practical and strategic needs. This allows for the distinction between two sets of planning needs. The planning methodology required to meet practical gender needs is essentially 'technical' in nature; it simply requires the development of the necessary tools and techniques to help women do better what they are already doing. In contrast, the planning methodology required to meet strategic gender needs is clearly 'political' in nature. Strategic gender needs derive out of the subordination of women to men in society. From the perspective of a political economy approach such redistribution of power and resources between men and women requires structural transformations.

Although the political economy approach has successfully incorporated conflict into planning, as a model it has been criticized as explaining rather than developing an appropriate planning process. Healey (1989) argues that *planning as debate*, a further recent planning methodology, provides the basis for a planning methodology with a political economy approach. This recognizes that planning is neither just a technical process, nor merely political, with its real concern the interaction between the two. The central premise of this model is that, in a plural society with a diversity of interests, often in conflict, the democratic mode of collective action has to proceed through debate.

> Planning as debate has the potential to confront those who 'currently' hold power at the level of ideology and philosophy as well as materially. It in effect changes the locus of exercise of power from the wielding of material resources, to association with a convincing argument.... Its challenge is to address rather than ignore tensions.
>
> (Healey 1989: 7)

Planning as debate provides useful guidelines for the elaboration of the gender planning process. For, unlike physical planning traditions with their clear-cut operational methodologies, obviously 'subordination' or 'emancipation' cannot be operationalized as such. The goal of gender planning refers ultimately to changes in the relationships between groups in societies, more specifically between men and women. Therefore, the planning methodology must place emphasis on planning as an iterative process. As Healey (1989) describes it, the notion of planning as debate assumes that outcomes in terms of values or strategies cannot be precisely anticipated. The premium, therefore, is on process, identified as the institutional mechanisms and operational procedures through which interests and needs are 'mediated' into strategies, policies, programmes and projects.

Gender planning is concerned with transformative processes that are highly political and can be assumed to involve conflict. This requires a planning methodology that emphasizes debate, negotiation and conflict resolution. The underlying assumption is that such a process may not necessarily be either rational or comprehensive. Planning as debate comes closest to this form of planning process. However, not only do the details of the methodology remain to be developed, but ultimately this may only happen through debate, decision and practice. Consequently, the following description of gender planning methodology can only be tentative and exploratory.

GENDER PLANNING: CHARACTERISTICS OF A NEW PLANNING TRADITION AND ITS METHODOLOGY

Where the planning tradition defines the objectives, purpose and agenda of gender planning, the planning methodology outlines both the planning process and the organizational structure that implements it. As a new planning tradition, the focus of gender planning, as shown in Table 5.3, is obviously on gender – that is, the social relations of inequality between men and women. The knowledge base for this new planning tradition comes from many sources. As discussed in Part One of the book, the most important are the feminist debates from both the First and Third World, and the more recent WID/GAD debates. The objective of this planning tradition is to achieve both strategic and practical gender needs. Its goal is to emancipate women from their subordination, and to embrace equality, equity and empowerment. Its agenda is the specific strategic gender needs of women, which, as outlined in Chapter 3, relate contextually to social relations such as class, ethnicity, 'race', religion and location. It is important to stress that the agenda of gender planning is not solely concerned with achieving practical gender needs of women. As the discussion on empowerment policy in Chapter 4 highlighted, practical gender needs are critical, but above all as instruments to achieve strategic gender needs.

Planning frameworks comprise several procedures. Ideally these identify a logical sequence of stages that describe the necessary number of actions required to complete the planning process. It would make gender planning easy to undertake if its planning process contained a number of finite stages of logical progression, such as the straightforward linear procedures of blueprint planning and rational comprehensive planning, illustrated in Tables 5.1 and 5.2. However, where the purpose is a change in attitudes rather than to produce products, solutions are not so simple. Whereas gender-aware planning may be able to adopt rational comprehensive planning procedures, gender planning cannot. As Huxley (1988), for instance, so aptly comments,

Table 5.3 Characteristics of gender planning as a new planning tradition and
methodology

Focus	Gender
Knowledge	Feminist theories and WID/GAD debates
Goal	Women's release from subordination and 'emancipation'
Objectives	Achievement of gender equity, equality and empowerment through practical and strategic gender needs
Agenda	Context specific strategic gender needs such as: – Equality in the gender division of labour within the household – Equal control over resources and power of decision-making within the family – Equality in the gender division of labour in paid employment – Equal participation in local- and national-level political processes
Planning framework	Iterative planning process utilizing: a) Gender planning tools b) Gender planning procedures c) Components of gender planning practice

'Neighbourhood houses and the design of communal kitchens will not, in and of themselves, change the social relations of patriarchy' (p. 42).

The focus of gender planning on social transformation means that its procedures relate to processes of negotiation and debate. These are concerned with the redistribution of power and resources within households, civil society, the state and the global system. The fact that there are very real problems in 'operationalizing' subordination means that the epistemology of this type of planning process is difficult to articulate. Also, much of the methodology relating to the procedures through which interests and needs are 'mediated' into strategies, policies, programmes and projects is context-specific. It cannot be determined a priori.

For reasons such as these, the gender planning process does not consist of a logical sequence of stages in an a priori defined process. It consists of an iterative process made up of a number of *procedures*, which constitute steps that are 'ongoing and overlapping' (Levy 1991: 7). Each of these procedures utilizes and incorporates a number of *methodological tools*, deriving out of gender planning principles. These iterative procedures, in turn, are integrated into the four different *components* of gender planning practice.

These three interlinked elements – methodological tools, gender planning procedures and components of gender planning practice – in total constitute the gender planning framework (see Table 5.3). The last section of this chapter outlines both the methodological tools and the gender planning procedures. The final four chapters then address the four different components of gender planning practice.

METHODOLOGICAL TOOLS

Part One of this book provided the conceptual rationale for the key principles of a gender planning methodology. In order to inform the manner in which planning is undertaken, these are translated into tools for planning practice, as illustrated in Table 5.4.[2] In gender planning as a planning tradition many of the more 'technical tools' easily identified in physical planning traditions do not exist. The goal of gender planning, to overcome subordination, is not easily susceptible to spatial and physical interventions. These refer to changes in the relationship between men and women. Consequently, many of these tools are essentially *performance indicators*, designed to measure changing processes rather than technical interventions. The following six principles can be identified, with their associated planning tools described.

Gender roles and gender roles identification

The first principle relates to *gender roles*. Existing gender divisions of labour manifest themselves through the different roles played by men and women. Obviously the concept of the triple role is a simplification of the complexities of the social construction of gender relations and divisions of labour in specific socio-economic contexts, and their changing dynamics over time. For planners, however, this provides the first key principle for a gender planning methodology that enables them to translate gender-awareness into a tool for planning practice.

Above all, *gender roles identification* is a tool that makes visible previously invisible work. Most women live in a situation in which only their productive work, by virtue of its exchange value, is valued as work. Reproductive and community managing work, because they are both seen as 'natural' and non-productive, are not valued. This has serious consequences for women. It means that often the majority, if not all, of the work that they undertake is invisible. Neither men in the community nor those planners whose job it is to assess different needs within low-income communities recognize it as work. In contrast, most men's work is valued, either directly through paid remuneration, or indirectly through status and political power. The purpose, therefore, of gender roles identification is not only to separate out the different tasks both men and women, and boys and girls do. It is also to ensure the equal valuing of these tasks through the identification of

Table 5.4 Gender planning principles, tools and procedures

No.	Principles	Tool
1	Gender roles	Gender roles identification
2	Gender needs	Gender needs assessment
3	Equal intra-household resource allocation	Disaggregated data at the household level
4	Balancing of roles	Intersectorally linked planning
5	Relationship between roles and needs	WID/GAD policy matrix
6	Equal control over decision-making in the political/planning domain	Gender participatory planning

cm = community managing
cp = community politics
p = productive
r = reproductive
GDOL = Gender Divisions of Labour
PGNs = Practical Gender Needs
SGNs = Strategic Order Needs

Table 5.4 (Continued)

Procedures	Techniques	Purpose
	Identification of p/r/cm/cp roles of men and women and equal allocation of resources for work done in these roles	To ensure equal value for women and men's work within the existing GDOL
Gender diagnosis, objectives and monitoring	Assessment of different practical and strategic gender needs	To assess those needs relating to male–female subordination
	Gender disaggregated data	To ensure identification of control over resources and power of decision-making within the household
	Mechanisms for intersectoral linkages between economic, social, spatial, development planning	To ensure better balancing of tasks within the existing gender division of labour
Gender entry strategy		
	Range of policy approaches: welfare; equity; anti-poverty; efficiency; empowerment	Performance indicator to measure how far interventions reach PGNs and SGNs
Gender consultation and participation	Mechanisms to incorporate women and representative gender-aware organizations into the planning process	Ensure SGNs are incorporated into the planning process

reproductive, productive, community managing and community politics roles.

Gender needs and gender needs assessment

The second principle relates to *gender needs*, and the assessment of those that are practical and strategic in nature. The planning tool for this is *gender needs assessment*. Its purpose is to recognize women as active participants in development (practical gender needs). It also recognizes that women do not participate in development on equal terms with men because of their subordinate position (strategic gender needs).

Gender needs assessment classifies planning interventions in terms of those that meet practical gender needs – that is, the needs identified to help women in their existing subordinate position in society – and strategic gender needs – namely, the needs identified to transform existing subordinate relationships between men and women. Because it deals directly with the issue of subordination, this planning tool is the crux of the framework for a gender planning methodology. As such, gender needs assessment is a tool by which it is possible to measure changes.

Intra-household resource allocation and disaggregated data at the household level

The third principle concerns *intra-household resource allocation* in terms of ensuring equal control over resources and power of decision-making between men and women within the household. The widespread popularity of the household as the unit of planning derives from the fact that it is analytically simpler as a unit of analysis for economic decision-making than is the individual. In addition, it is much easier for planners to direct resource flows and benefits to the household as a unit. But the welfare of family members cannot be read off from the socio-economic characteristics and economic choices made by the household head. Neither does the distribution of resources at the household level guarantee that the benefits will trickle down. Assessments of poverty based only on household income may thus conceal much hidden poverty. Equally, resource allocation policy focused at household level does not necessarily deliver benefits to all members of the household. Women and children are often chronically under-resourced relative to men in many poor households. Once intra-household allocation is viewed as the outcome of bargaining processes, then planners can recognize that the individual to whom the resources are directed most frequently decides the control over allocation. Gender-based responsibilities in most societies show that it is more efficient to allocate benefits directly to women if children's welfare is the main priority.

The necessary planning tool here is *disaggregated data at the intra-household level*. Its purpose is to ensure that planning equally benefits all members of the household by allowing women access to and control over resource allocations. It will, therefore, ensure that planning interventions benefit women by reaching their practical gender needs.

Balancing of roles and intersectoral linked planning

The fourth principle relates to the balancing of roles, which derives from gender roles identification. For women, this relates to the co-ordination of their triple role. Women experience competing demands between reproductive, productive and consumption activities. Productive work has to fit round family and community-level responsibilities. It is therefore often part-time or carried out at home. For women the key problem is the relationship between wage, agricultural, child-care or community managing work. The fact that the burden of simultaneously balancing these roles severely constrains women is ignored. Hence, traditional sectoral-based planning is often unhelpful for women. Planning which focuses, for example, on provision of transport or services, or any sector on its own, does not consider links with other sectors. Thus it does not realize the constraints influencing women's use of goods and services provided by planning initiatives.

There is a need, therefore, for *intersectoral linked planning* that links different activities and scales of planning, such as home and transport or workplace and environment. This will ensure that the goods and services provided can be utilized by women to balance their tasks better within the existing gender divisions of labour.

The relationship between roles and needs and the WID/GAD policy matrix

The fifth principle relates to the *relationship between roles and needs*. The five different policy approaches to WID that have been identified all differ in terms of the triple roles of women that they recognize. They also differ in terms of the practical or strategic gender needs they meet, and the extent to which they include participatory planning procedures. This provides the conceptual rationale for a further principle of gender planning, which relates to the relationship between roles, needs and planning processes, and the fact that meeting strategic gender needs changes existing gender roles. The relevant planning tool is the *WID/GAD policy matrix*. This is a performance indicator to measure the extent to which different planning interventions transform the subordinate position of women by meeting both practical and strategic gender needs.

Equality and incorporation in the planning process

The sixth principle relates to *equality between men and women* in the planning process. While the participation of local women in the planning process may result in greater control over the allocation of specific resources, it may not reduce intra-household inequalities. There are significant constraints arising from different perceptions of needs between men and women, and this has implications in terms of intra-household entitlements. These constraints suggest that changes in women's outlook and increases in their freedom within and beyond the family unit require further planning principles and planning tools.

While access to productive work can help women, changes in household dynamics are often more influenced by extra-household experiences of collective action. In this respect women's solidarity groups play a critical role in providing women with the space and opportunity to question their subordinate status. Such groups empower them to confront and transform the oppressive aspects of intra-family and household relations. The planning tool for this purpose is the *incorporation of women, gender-aware organizations and planners into planning*. This aims to ensure that real, as against perceived, practical and strategic gender needs are identified and incorporated into planning processes.

These tools of gender planning challenge current planning stereotypes. They can be utilized in planning practice to help planners appraise and evaluate ongoing planning interventions. Tools such as the triple role, gender needs assessment, the WID/GAD matrix and gendered participatory planning can also help them to formulate and implement gendered proposals at policy, programme or project level.

GENDER PLANNING PROCEDURES

Methodological tools such as these are not only important for planning generally. In addition, they are fundamental inputs for incorporation into planning procedures specific to the gender planning framework. Following Levy, these are identified as gender diagnosis, gender objectives, gender monitoring, gender consultation and participation, and gender entry strategy (Moser and Levy 1986; Levy 1991). As mentioned above, these procedures are not stages in a logical planning framework, but are iterative overlapping procedures that can be incorporated at any stage of the planning process. It is important to describe each in turn.

Gender diagnosis

As its name implies, gender diagnosis is concerned to identify the particular implications of contextually specific problems of development *for men and women*, and the relationship between them. Levy identifies two stages in such diagnosis. First, analysis of the problem using methodological tools such as gender roles identification, gender needs assessment and the WID/GAD policy matrix. The second stage is the 'organization of problems into a cause and effect hierarchy to establish a gender aware perspective on the dominant problems' (Levy n.d.). She comments that gender diagnosis is an ongoing activity that is carried out at all key points of an organization's planning cycle.

In undertaking gender diagnosis at the planning stage it is important to identify the socio-economic information on gender roles already available, and what needs collecting. At a minimum, gender diagnosis usually requires data on the following issues: the division of labour in productive activities; the division of labour, by age and sex within the household, including seasonal difference; the structure and size of households, and stage in the life cycle; sources of household income, including off-farm activities; control and decision-making within the household over cash, land and other resources; the structure and composition by age and sex of community-level decision-making bodies; local- and national-level political structures.

Gender diagnosis integrates gender planning tools such as gender roles identification and disaggregated data at the household level (see Table 5.4). The use of such tools often requires the collection of accurate gendered information. Implicitly, such critical questions as how 'objective' are the data, who collects the data, and who has access to the data have to be addressed. Obviously, all planning methodologies require that planners have accurate and reliable information. However, if people are to be involved in decision-making and debate throughout the planning process, they also require access to information, and this is not always so obvious. Lack of access to information disempowers communities involved in decision-making processes.

The assumption that data are objective and 'value-free' is particularly problematic for a planning tradition concerned with transformation, and the redistribution of power and resources within society. So much of the bias in data collection has been the result of the very biases of researchers. This makes it important for women to participate in the collection of data concerned with gender issues, for the manner in which questions are asked, and the sex of the researcher can decide research outcomes. In many societies, unless women speak, data will only reflect men's view of the world. Therefore, a precondition of gender diagnosis is the involvement of women researchers in data collection. In addition, it includes consultation with community women and with local, progressive, gender-aware organizations, in which women themselves are the source of this information.

Within specific planning contexts gender diagnosis questions current statistical research stereotypes about the household as the unit of analysis, and the male breadwinner as the household head. Both have had important implications for the 'invisibility of women'. One of its objectives, therefore, is to identify the gender biases in data analysis. This requires the identification of the roles and needs of both women and men in the household and the community, as well as difference in household structures and in intra-household control over resources and power.

Among the extensive number of questions addressed are those concerning the assumptions made about the roles of men and women in society. Are these explicit? Are they realistic? Within their roles, across societies, men and women can be involved in different activities. For example, in Asia unskilled construction work is accepted as 'women's work', while in Africa or Latin America this is identified as 'men's work'. Since gender relations are socially constructed, they are contextually specific and often change in response to altering economic circumstances, and consequently cannot be read off check-lists.

In the second stage of gender diagnosis, where cause and effect are identified, the most critical tool is often *gender needs assessment*; that is, the gender-specific understanding of what is causing subordination and the barriers preventing its elimination. It is this that leads into the identification of gender objectives.

Gender objectives

As defined by Levy, these are derived from the definition of the dominant problems in the gender diagnosis. 'They are a guide to action at particular points in the process of making policies, programmes and projects more gender aware' (Levy n.d.). Gender objectives are redefined as the process proceeds, responding to changes brought about either by actions to make interventions more gender-aware, or by other factors external to the process.

Gender objectives provide the basis for a specific agenda that identifies which gender needs are to be selected, and the strategies to accomplish this. Gender objectives, like gender diagnosis, then become an iterative process leading to the re-identification, adjustment and refinement of practical and strategic gender needs, thus ensuring that they can be introduced into any stage of the planning process.

Gender monitoring

Gender diagnosis provides the performance criteria to appraise and evaluate the extent to which actions and interventions achieve gender objectives. Again these are iterative processes, used at different stages in any planning cycle to assist in the redefinition of gender objectives. Monitoring and evaluation can be undertaken with specific questions using gender planning tools. The following example shows how one such tool, gender roles identification, can be used in monitoring.

(a) What will be the impact on women in their productive role?

Will the programme/project positively or negatively affect women's access to

- land, particularly rural land for food crops for household and market consumption and urban land for housing?
- opportunities for paid employment or other income-earning activities, particularly when women's existing sources of income are destroyed or reduced?
- the labour of other household members for economic activities existing or new technology?
- credit, particularly where formal collateral arrangements for women do not exist?
- skill training and information?
- income generated from her productive work?
- basic services such as transport, water and fuel?

(b) What will be the impact on women in their reproductive role?

Will the programme/project positively or negatively affect women's access to

- the labour of other household members for domestic work?
- items of household consumption such as firewood or water?
- basic services such as health clinics, transport?
- food of adequate nutritional levels?
- land for housing?
- skill training and information?
- income generated for the household: will it result in greater or lesser dependence on men's cash income for household food and necessities?
- labour-saving technology such as handmills for food preparation?

(c) What will be the impact on women in their community managing role?

Will the programme/project positively or negatively affect women's access to

- existing community-level decision-making?
- new community-level decision-making introduced by the project, if relevant?
- income, particularly where the project relies on their voluntary labour?

(d) What will be the impact on women's ability to balance their triple role?

Will the programme/project increase woman's work in one of her roles, to the detriment of her other roles?

- if increased time is required in productive work – for example, weeding of land for new cash-crop production – will this reduce her capacity to undertake existing tasks in her triple role?
- if increased time is required in reproductive work – for example, increased distance to collect firewood – will this reduce her capacity to undertake existing tasks in her triple role?
- if increased time is required in community managing, such as voluntary labour to maintain project infrastructure, will this reduce her capacity to undertake existing tasks in her triple role?

Gendered consultation and participation

In a planning methodology that emphasizes debate, negotiation and conflict resolution, gendered consultation and participation is the most critical yet complex gender planning procedure. Although theoretically 'participation' in planning is generally considered a 'good thing', clear consensus still does not exist as to what it means in practice – and throughout the ensuing text the term 'participation' is used generically to include consultation. A large number of definitions reflects the ideological range of interpretations of development and different approaches to planning, which have themselves changed over time.[3] The more the rhetoric of participatory planning is evoked, the greater is the diversity of interpretations as to what it means. The distinction between the four questions of why participation, when to participate, whose participation and how to participate provides a useful framework for understanding some complexities of this issue. These are identified generally as well as specifically in terms of gendered participation in planning.[4]

(a) Why gendered participation?

This question refers, obviously, to the reasons for and causes of participation. One useful general definition of the objectives of participation in development projects is that of Paul (1987), who identifies a fivefold continuum. This ranges from the objective of participation for empowerment, to capacity-building, through increasing project effectiveness, to improving project efficiency and, finally, to project cost-sharing. Central to the debate about participation is the issue whether participation includes an element of empowerment. Here the distinction between 'means' and 'ends' is useful (Oakley and Marsden 1984). Where participation is a means, it generally becomes a form of mobilization to get things done. This can equally be state-directed, top-down mobilization, sometimes enforced, to achieve specific objectives, or bottom-up 'voluntary' community-based mobilization to obtain a larger share of resources. Where participation is identified as an end in itself, the objective is not a fixed, quantified development goal, but a process, the outcome of which is an increasingly 'meaningful' participation in the development process (UNRISD 1979; Moser 1983, 1989b).

This is particularly significant for gendered participation and the active consultation of women in the development process. Projects frequently rely on women's participation as a means to ensure project success. However, they far less frequently recognize that for women as much as for other groups in society, participation is an end in itself. In highlighting this issue for planners involved in housing and human settlement projects, I advocated three reasons, each with a different underlying objective, for incorporating women's participation. The first objective identified below relates to empowerment; the second concerns participation as a means to achieve efficiency, effectiveness and cost recovery; the third objective of woman's participation focuses on capacity-building.

a) Women's participation is an end in itself. Women as much as men have the right and duty to participate in the execution of projects which profoundly affect their lives. Since women accept primary responsibility for child-bearing and rearing, they are affected most by housing and settlement projects. They should, therefore, be involved in the planning and decision-making as well as the implementation and management of projects that relate particularly to their lives.

b) Women's participation is a means to improve project results. Since women have particular responsibility for the welfare of the household, they are more aware than men of the needs for infrastructure and services and are also more committed to the success of a project that improves living conditions. The exclusion of women can negatively affect the outcome of a project, while their active involvement can often help its success.

c) Participation in housing activities stimulates women's participation in other spheres of life. Through active involvement in housing projects women may be encouraged to participate fully in the community. Participation in projects has been seen as an important mechanism to 'overcome apathy' and 'lack of confidence' and it can make women visible in the community. It can enable women to come out of their houses and show them the potential of self-help solutions. In so doing, it may raise awareness that women can play an important role in solving problems in the community.

(UNCHS 1986: 2)

(b) When to do gendered participation

This refers to the different stages or phases involved in a policy, programme or project. It refers not only to the gender analysis stage, described above, but also the whole planning process. In project planning, for instance, this is most commonly identified as decision-making, implementation, financing and managing. However, it also includes a number of other phases depending on the project type.

The extent to which planners block or adopt a gender planning methodology depends on whether there is participation by women in the identification phase when general policy directions and goals are decided. Much of the planning for WID, for instance, totally ignores this stage, but this is when it is necessary to discuss the nature of gender subordination in the prevailing system, or structure of society. This sets the framework for what is essentially a highly political process.

The basis of such a framework is the distinction between whether the issue identified is one of 'gender' or women. Is the issue seen as strategic, relating to the fact that women are subordinated? Or is it decided as practical, relating to female marginality due, for example, to lower educational levels or unequal access to employment? This distinction determines whether a gender planning or gender-aware planning approach will be adopted. It is the former planning approach that specifies its general goal as the *emancipation* of women.

(c) Whose gendered participation?

In identifying who gains access to the debate about the gender planning process and on what terms, planners have a responsibility to argue for a broad-based approach including different interests and attitudes (Healey

1989). Once the notion of a single public interest or a coherent political consensus disappears, an inclusionist approach must be advocated. For any participatory planning process this also includes identifying who *controls* the process of the identification of goals and objectives, and who legitimizes the process. Whose judgements are used about what is necessary or satisfactory? These questions refer not only to whether it is the planners or the people, but also whether it is men and women, local communities as much as elite groups.

The tendency is to understand 'the community' in homogeneous terms, with further disaggregation only done at the household level. This means that the policy level rarely mentions the important role women play in participatory planning. It occurs despite the constant comments from those working on the ground that 'without the women the project would have failed'. The fact that men are more likely to be involved in community politics means that the participation of local women as community managers is frequently either invisible or not valued. However, there is also a negative side to women's participation. While their participation is often crucial for project success, this is based on the assumption that women have 'free time'. Lack of awareness of women's triple role, therefore, can be a cause of project failure. When women fail to participate, it is not women who are the problem, as frequently identified. It is a lack of gender-awareness of planners about the different roles of men and women in society and the fact that women have to balance their time allocation in terms of their three roles.

(d) How to do gendered participation?

This refers to the mechanisms by which gendered participation is accomplished. It addresses the question of how the dynamics of the participatory process resolve themselves in reality. Here the contradictions between intentions on paper (often lip-service) and the real agenda (often hidden in the planning stage) can become apparent in the practice of gendered participation.

The question of how to do gendered participation focuses directly on the style and procedures of the gender planning debate. For instance, if the decision-making stage includes participation, the methods used and the timing of consultation can influence the target groups reached. In many low-income programmes and projects, participatory methods of recruiting beneficiaries often inadvertently discriminate against or exclude women. This is true of scheme announcement methods, applications procedures and down-payment requirements. When authorities assume that everyone reads newspapers or public notices, with information distributed in written form, they often miss women. In many parts of the world women's access to education is considerably less than that of men. In addition, women are less likely to be exposed to information because of their lack of daily mobility. Even where planners use more direct methods of advertising, such as meeting where eligible applicants are likely to live, domestic responsibilities often prevent women from attending. If they attend mixed meetings they generally stand at the back, on the assumption that they will not talk. Real consultation with women often demands house-to-house consultation or small-group, all-women gatherings rather than large-scale meetings.

The capacity of women to be involved in such participatory processes requires changes in the ways in which participatory consultation is undertaken. These relate not only to such logistical problems as the hours of meetings; it also includes more fundamental problems relating to female consciousness and its articulation in participatory processes. Where negotiation and debate are critical to the planning process, judgements are made as to what values should govern decisions (Healey 1989). One important issue relates to differences between men and women in their perceptions of personal welfare and self-interest, against family welfare.[5] Sen (1990) distinguishes between well-being and agency, and the fact that a person may have various goals and objectives other than the pursuit of his or her well-being. He also identifies gender differences in perceived contributions to general family prosperity. These result in bargaining disadvantages that feed on themselves generationally through feedback transmission. Thus he comments that 'The political agency of women may be particularly important in encountering the pervasive perception biases that contribute to the neglect of women's needs and claims' (1990: 149; see Chapter 3 for more details). He therefore identifies the importance of processes of politicization – including a political recognition of gender issues – causing sharp changes in these perceptions. The role that women's organizations play in addressing this issue will be further examined in Chapter 9.

Participation presupposes a pro-active capacity and willingness to negotiate and debate throughout the planning process. Because of the way that women are so effectively excluded from real decision-making, in reality they often chose to withdraw rather than participate in planning processes. A useful concept that addresses this aspect of gendered participation is the analytical distinction between *'voice'* and *'exit'* (see Paul 1991).[6] The capacity to exert voice (pressure to perform) has received far more attention in the literature on gendered participation than has the capacity to find *exit* (alternative sources of supply). However, they may be equally important in the planning process.

Gender entry strategy

Gender diagnosis results in an understanding of the mechanisms of gender bias in terms of the underlying assumptions of sectoral policies and organizational structures. Gendered participation ensures negotiation and debate in the planning process. Gender objectives set the agenda for intervention. None of these procedures in themselves, however, can ensure that interests and needs are mediated into policies, programmes or projects. Gender planning is inherently a contextually specific political activity. Therefore it requires an entry strategy. As Levy has argued, an entry strategy is a

> prioritized and tactical set of actions designed to expand the room-for-manoeuvre at a particular socio-economic and political juncture to overcome constraints which may block or subvert desired gender interventions, and to utilize potential which may provoke a resource or an opportunity to promote them.
>
> (Levy n.d.)

Essentially, a gender entry strategy defines what it is possible for gender planning to achieve in a specific context. Entry strategies can be identified as having two phases. First, gender objectives, deriving from gender diagnosis and gendered consultation and participation, can be used to identify the critical points where practical gender needs have the capacity to reach strategic gender needs. The second stage is an assessment of the constraints and opportunities provided by institutional structures and their operational procedures. This is to ensure that the planning agenda can be translated into practice. The identification of entry points involves strategic political choices. Since those involved in the process influence its outcome, gendered participation and the consultation of women is important throughout the planning process.

Table 5.5, which identifies the three interlinked elements in a gender planning framework, can only illustrate one stage in the iterative planning process.

COMPONENTS OF GENDER PLANNING PRACTICE

In the past decade gender policy and gendered development policy have made impressive advances. In both cases gender diagnosis and gendered consultation and participation have been important. But WID policies frequently fail to become implemented practice, while the content of WID/GAD policies often changes during the implementation process. The increasing priority for those working on gender issues relates to *the inability to translate formulated gender policy into practice*. This is not an easy task. In the complex reality of planning processes it is often difficult to separate out the interrelated factors that influence and decide the implementation process.

Table 5.5 Planning methodology 3: one stage in an iterative gender planning
framework

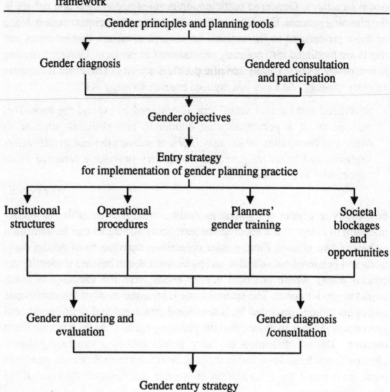

Nevertheless, if changes are to be made in current procedures it is critical to
identify the constraints and opportunities acting upon them.

The development and adoption of gender planning as a new planning
discipline with its particular principles, tools and procedures does not ensure
its successful implementation into planning practice. It does not mean that
current planning practices operationalize them, that existing organizational
structures institutionalize them, or that professionals involved in planning
make *gender-aware decisions*. Part Two of this book, therefore, focuses on
the implementation of four components of gender planning practice. The first
three are identified as the institutional structures, operational procedures and
behaviour of planners involved in gender planning practice. To address such
issues answers several important questions.

Chapter 6 focuses on the constraints and opportunities in the institutionalization of gender planning in a diversity of administrative and

organizational structures concerned with development-planning issues. The major question addressed is whether gender planning is more successfully institutionalized within existing mainstream planning organizations or within specially formed organizational structures. Chapter 7 then examines operational procedures necessary not only to formulate gender policy, but also to ensure its translation into practice through programmes and projects. Are existing methodologies adequate to operationalize gender policy, or is a separate planning methodology required? Chapter 8 turns to the issue of training to evaluate its effectiveness in overcoming the constraint of gender-blindness among professionals concerned with WID and GAD issues. Is the distinction between the gender-aware planner and a gender planner valid? Finally, Chapter 9 addresses the fourth very different component of gender planning practice. This concerns the organizational structures in society effective in blocking, or providing opportunities for, gender planning. It examines the conflicting, often competing, interests of different social groups. These frequently determine both the content and the context of planning. Ultimately, they also decide the extent to which women's groups, feminist organizations and non-governmental organizations are able to influence gender planning through both 'voice' and 'exit'. The question posed in the final chapter, therefore, concerns the extent to which 'bottom-up' alternative organizations outside government in fact provide the best entry point for gender planning.

6 The institutionalization of gender planning

When a new issue, such as gender, appears on the planning agenda, the first question raised is who will deal with it? Is it necessary to create an entirely new institutional structure or is it more appropriate to institutionalize it within existing mainstream organizations? Or is the best strategy simultaneously to do both? Underlying these different alternatives is the fundamental issue whether gender can be 'grafted' successfully onto existing structures, or whether it requires other structures to be effectively integrated into planning. What is clear, however, is that recognition of gender as a policy concern has not resulted in its automatic institutionalization into the wide range of agencies implementing policies for low-income communities in the Third World. Therefore, in the implementation of gender planning practice the first constraint for consideration is that of institutional factors.

In the past decade, organizations concerned with incorporating WID/GAD policy into planning and practice have identified the lack of adequate administrative structure as their most important constraint. The strategy frequently adopted has been the creation of new structures. Their specific mandate then is to translate gender policy into planning and practice. Globally, a proliferation of new organizations has emerged. These include ministries of women's affairs in national government, gender units in NGOs, and women's divisions in donor agencies. Only a few organizations have specifically resisted creating separate designated organizational structures and chosen to integrate WID/GAD entirely through 'mainstreaming' it within existing structures.

This chapter examines the institutionalization of gender and women and development policy, identifying constraints and opportunities in current strategies. It provides a brief historical background before describing the comparative strategies of several institutions currently integrating a WID/GAD policy within their institutions during the 1980s. These include national governments such as the Caribbean CARICOM countries. In addition, it includes bilateral donors, such as the UK Overseas Development

Administration (ODA) and the Swedish Development Agency (SIDA). Finally, it also refers to multilateral donors, such as the World Bank and the Food and Agriculture Organization (FAO), and UK-based development NGOs such as Christian Aid and Oxfam.[1]

TECHNICAL AND POLITICAL CONSTRAINTS IN THE INSTITUTIONALIZATION OF WID/GAD

In such an analysis it is important to highlight the extent to which the institutionalization of WID/GAD is simply a technical problem relating to existing institutional structures, or whether it is a political problem requiring change in the nature of institutions themselves. This raises two fundamental questions concerning the meaning of institutionalization. First, is it possible to 'gender' an institution? Second, is the purpose to create a 'gender-aware' organization, or a 'gendered organization'? The first question has implications for the relative success of the two different organizational strategies, that of creating new, separate institutions or of mainstreaming within existing structures. Whereas the tendency in published research has been to focus on the technical nature of institutional components, Staudt (1983) has pointed out the essentially political nature of institutionalization. According to the 'bureaucratic politics' model, organizational action is a result of the bargaining that occurs between hierarchically positioned players in an organization. The probability of success in bureaucratic games depends on bargaining advantages, skill and will in using resources. Prospects for effective bargaining depend on power resources, which include expertise, control over material resources, structural location and internal alliance building.

Despite a common concern to institutionalize WID/GAD, not only do organizational strategies often differ, but so also does the mandate as to what 'institutionalizing' effectively involves. This raises a second fundamental question. Can it be assumed that an institution that is gender-aware (because all those working within the institution recognize the importance of gender) will act in a 'gendered' manner? Two issues are relevant here. First, although an institution may be gender-aware, it may nevertheless still be structured as before. To have a gendered institution, as many feminists have argued, it may be necessary a priori to change the institutional structure. Secondly, it cannot be assumed that a gendered institution will act in a neutral, objective manner. This assumes that once technically perceived problems of lines of responsibility and gender-awareness are ironed out, successful implementation can occur. Do organizations operate 'objectively', such that institutionalized interventions are 'neutral', in effect, or are decisions made in the interests of particular groups?

Staudt (1983) has argued that bureaucratic resistance to women's programmes may be greater than the usual resistance to new mandates. In her analysis of the 'bureaucratic mire', she maintains that institutions are political actors in their own right. Consequently, bureaucratic resistance must be seen in terms of the fact that the prospect of female empowerment threatens male privilege (1990). The creation of new institutional structures, or the gendering of existing structures, may not in themselves ensure success-ful implementation. From the outset it is necessary to recognize that political constraints and opportunities may determine the extent to which the institu-tionalization of WID/GAD makes any fundamental changes in the planning process.

HISTORICAL BACKGROUND: THE EXPERIENCE TO DATE

In the history of planning, new organizational structures commonly develop for the institutionalization of new planning procedures. Once established, such organizations very rapidly tend to become the legitimized 'home' of a particular sectoral issue or planning concern. This makes it difficult for them to absorb new issues. It also means that once a planning concern is identified with a particular administrative structure it is difficult for other organizations to institutionalize it.

For these reasons, the recent institutionalization of WID/GAD must be viewed in terms of its historical origins. As described in Chapter 4, colonial and post-colonial governments first identified the problems of Third World low-income women as a policy concern. Residual welfare social policy during this period considered women, along with other 'vulnerable groups', as a marginal issue. If they institutionalized such policy at all, it was into weak, under-financed ministries of social welfare. More frequently, how-ever, welfarist, voluntary-based, local NGOs picked it up.

Prevailing macro-economic and social policy has resulted in the institu-tionalization of low-income women's needs into peripheral and under-resourced organizations. In addition, it is the result of the difficulty experienced in locating women, a category of people with cross-sectoral importance, within planning machineries based on sectoral priorities. Despite the widely acknowledged problems this poses for non-sectoral concerns, most planning procedures, particularly those at government level, continue to plan with a sectoral approach. The problem, thus, is not unique to WID, but has also dogged other cross-sectoral concerns such as community devel-opment, rural development and, most recently, the environment. Despite the obvious lack of analytical logic, these are frequently identified as sectoral concerns (Moser 1989b; Conyers 1982). One successful institution to cope with cross-sectoral planning is UNICEF. Their mandate is to plan for children, although interestingly enough over time they have extended this limited target to include women and others concerned with children's welfare (Moser 1989b).

The institutionalization of WID concerns into welfarist, under-resourced organizations, whether at national, international or donor level, was a legacy inherited by the UN Decade for Women. Until 1975 most Third World countries had not addressed women's concerns in their own right. Yet, as early as 1962 the UN Commission on the Status of Women first identified the value of appointing national commissions on the status of women. These made recommendations for improving the position of women in their respective countries (UNCSDHA 1987). Subsequent UN meetings focused attention on institutional mechanisms as an important part of national strategies for the advancement of women. However, it was not until the first major conference of the UN Decade of Women, in 1975, that the endorsed World Plan of Action identified national machinery in the following terms:

> The establishment of interdisciplinary and multisectoral machinery within government, such as national commissions, women's bureaux and other bodies, with adequate staff and budget, can be an effective transitional measure for accelerating the achievement of equal opportunities for women and their full integration into national life. The membership of such bodies should include both women and men, representative of all groups of society responsible for making, and implementing policy decisions in the public sector.
>
> (UN 1976a: para. 34)

It recognized the cross-sectoral nature of the concern and identified that representation should include government ministries and departments responsible for education, health, labour, justice, communications and information, culture, industry, trade, agriculture, rural development, social welfare, finance and planning.

During the 1976–85 Decade of Women, 'machineries' were more clearly defined. This involved a shift from the early emphasis on advisory bodies to a broader idea of machinery in multiple and more complex forms, intended to emphasize official institutionalization within government. Its popularity with international donors has been explained in the following terms:

> At the time it was also believed, it seems, that WID belonged to that group of trendy issues that the UN introduces from time to time to stimulate the imagination of the international community to strengthen its sense of solidarity with less fortunate members, and to prevent opinion leaders from losing interest in matters of global concern.
>
> (Himmelstrand 1990: 104)

By 1985 most countries identified some form of national machinery, 50 percent of which they set up during the decade. Of the 127 member states listed, approximately 22 percent had organizations for women located in ministries of social affairs or social welfare. A further 20 percent were in NGOs, about 16 percent in separate ministries, and about 17 percent located in the office of the prime minister (UNCSDHA 1987).

More recently, bilateral donors have also addressed the institutionalization of WID. All nineteen members of the OECD's Development Assistance Committee (DAC) have adjusted to facilitate the implementation of their WID policies. Centralized WID Units work as catalysts for WID issues within the organizations, giving WID assignments to designated officers in the various departments/divisions of the agency concerned. While each agency has the minimum of one permanent WID adviser, twelve have WID Units. These comprise several professionals and support staff, located in strategic positions to provide WID expertise to those departments or divisions involved in policy-planning and operations (OECD 1990). A recent survey of European NGOs working in the South showed that 31 percent had a person, group or department dealing with gender issues. This figure ranged from 4 percent of French NGOs to 61 percent in the Netherlands, many of whom are part-time or volunteers (LCD 1989).

Despite such an array of new administrative structures, clearly women's 'machineries' have not fulfilled their expectations. Even by 1985 the UN, one of its greatest proponents, had expressed doubt concerning the efficacy of the mechanism in place. National machineries were expected to have a multiple function, with both a planning and advocacy role, carrying out decisions and playing a catalytic role. The UN, therefore, concluded that 'Looking back it is clear that too much was expected too soon from widely disparate bodies, in very different locations' (UNCSDHA 1987: 7).

An INSTRAW assessment of the 'failure' of women's machinery in seventy-nine countries identified a number of factors responsible. These included small budgets and staff, attitudes that legitimize female subordination, mandates that focused on welfare, under-funding of women's projects and separate women's divisions, and ministries lacking the resources and time to influence other government ministries (cited by Staudt 1990). Similarly, Northern NGOs list lack of resources, internal decision-making structures, internal staffing policies, insufficient awareness of gender issues, and lack of willingness to change as the most frequently encountered problems (LCD 1989). Finally, the OECD (1990) reported that limited staff resources have been a major constraint against the adoption of WID policies.

Staudt provides additional evidence to support the assertion that such machineries have 'failed', citing the budgetary allocations of multilateral and bilateral donor agencies. In the UN agencies, less than 4 percent of project budget allocation benefit women, with a cursory 0.2 percent of budget allocated directly to them. Less than 1 percent of FAO projects contain strategies to reach women. Finally, of the $700 million budget of the UN Development Program, UNDP, UNIFEM receives only $5 million (Staudt 1990). Nor is the situation of Northern NGOs working in the South more encouraging. One 1988 survey showed that 93 percent of European NGOs did not have a separate budget line for women's projects. Of those which did, the women's budget represented an average of 9 percent of total budget (LCD 1989).

It is now widely agreed that the creation of women's 'machineries' in governments, donor agencies and NGOs raised expectations that women's needs would be addressed. It also manipulatively gave the impression that the creation of separate institutional structures met such needs. This has often resulted in a total reduction in the resources reaching women. Thus Staudt, writing in 1990, concludes that the future looks bleak:

> At least at the aggregate level, little progress has been made in dismantling institutionalized male privilege. Such institutionalized privilege is deeply embedded in the state and grounded in western cultural heritage that has spread throughout the world in different degrees. Prospects for redistributive change look grim.
>
> (1990: 3)

WHAT HAS GONE WRONG? CONSTRAINTS IN CURRENT INSTITUTIONAL STRATEGIES

In retrospect, was it naïve to assume that the creation of an organizational structure by itself could successfully and effectively institutionalize a new preoccupation on the planning agenda? Can the creation of a new organization, a product as it were, introduce overnight a process, that of creating a gendered institution? Certainly the UN recognized that the original 1975 definition was too ambitious when re-endorsing national machineries in the 1985 Forward Looking Strategies. The Nairobi World Conference to Review and Appraise the Achievements of the Decade adopted a more modest definition. It agreed only that appropriate government machinery for monitoring and improving the status of women should be established where it was lacking. It concluded that 'To be effective this machinery should be established at a high level of government, and should be ensured adequate resources, commitment and authority' (UN 1986: para. 57).

Given the short time that has elapsed since their establishment, is the widespread dismissal of women's machineries premature and unjustifiable, displaying an unconstructive level of impatience? This apparent lack of progress calls for a re-assessment and more rigorous understanding of the underlying causes of institutional failure.

Is 'failure' the result of technical constraints relating to institutional structure, or are there more fundamental political constraints at work? Has a conflation of the two resulted in an unnecessary level of disillusionment? This is particularly destructive for those working within such organizations, on a day-to-day basis, at the coal face. Does it dissuade them from pushing forward the debate? An accurate understanding of the constraints and opportunities in institutionalizing WID may be important for such employees. With a decade of experience, it is imperative to examine the most important constraints identified and the way different institutions, successfully or unsuccessfully, have tackled them. Ultimately this will assist in a redefinition of the criteria by which succcess and failure are measured.

Is a separate organization or unit the problem? The experience of Northern NGOs

Most governments, donor agencies and NGOs have set up separate organizational structures to institutionalize WID. The first question concerns whether this marginalizes the issue. Is it an appropriate institutional strategy to mainstream within existing structures? As Anderson and Chen (1988) have identified, the underlying rationale for the two strategies differs. One strategy establishes specially designated WID offices, whose responsibility it is to raise the issues of WID for the institution as a whole, and to carry out WID programmes

> on the assumption that the compelling argument for the centrality of women to systematic development is not, and will not be self-evident; someone must take on the advocacy, teaching and watch-dogging roles if WID is to be taken seriously and if institutional resistance is to be overcome.
>
> (Anderson and Chen 1988: 2)

This strategy requires gender specialists to take responsibility for the institutionalization of WID, from a separate institutional structure.

By contrast, another approach prioritizes the integration, or mainstream, of WID within existing institutional structures, into established programme areas and sectoral activities. It argues that institutional responses will only be significant and convincing if, and when, the core staff and programmes of the organization accept and integrate WID. Institutionalization will be more effective if carried out by gender-aware generalists from their particular positions within the organization.

The comparative experience of two Northern international NGOs based in the United Kingdom and funding Third World development projects highlights some of the advantages and disadvantages of these two alternative strategies. Oxfam is one of the largest UK NGOs with a staff of more than 1,000. It has created a designated WID office, the Gender and Development Unit (GADU), emphasizing the importance of gender specialists. In contrast, Christian Aid, a smaller organization numbering about 230 staff, has chosen a mainstream approach. In Oxfam the impetus to create a special WID focus came 'bottom-up', from a group of women field officers. Working in collaboration with regional and headquarters staff, they proposed the establishment of a separate unit (Williams 1983). In Christian Aid, its Women's Group (an advisory body of women working in the organization) drew up a set of organizational recommendations in 1983. These identified constraints and blockages to WID. In 1987 a second set followed, when the group formalized itself into a Women's Forum (Christian Aid 1987). Thus, in common with most organizations, in both cases the original catalyst came in the early 1980s – 'bottom-up', and from women within the organization.

GADU was established in Oxfam in 1985 as one of the first initiatives of a new overseas director, with its purpose 'to tell people about the gender issue'. He perceived this as a short-term initiative that would cease once everyone had understood the issue. In Christian Aid one of the divisions, the Aid Sector, responded to the Women's Group's recommendations. This came from a concern that the organization's project documents did not adequately incorporate gender issues. In Oxfam the final impetus for the creation of a separate institutional structure was a top-down directive from a sympathetic senior male staff member. In Christian Aid the planning pre-occupation of its division directly concerned with the allocation of project finance was particularly important in endorsing the initiative to mainstream gender. At its inception, Christian Aid's support for a WID initiative had a broader support base than that of Oxfam, where one senior-level staff member played a deciding role. Different policy approaches in these two NGOs, therefore, influenced the particular procedures and methods used to institutionalize gender.

Different organizational strategies were also important. In Oxfam GADU was one of several specialist units which also included research and evaluation, technical and health units. In a management structure which excluded them from the main line, their function was an advisory one. Because Oxfam defines itself as a 'field-led' organization, with most of its staff working in the Third World, GADU, like other units, had a 'responsive approach'. It started by focusing its work at the field level, responding to requests for advice and funding. This was followed by the promotion of the employment of gender project officers in the field to do the work '*in situ*', and more recently the endorsement of the need for regional offices to develop a gender policy. It identified networking through regional workshops as a means to influence the attitudes and working practices of staff in their Oxford headquarters (Oxfam, n.d.). Given its advisory mandate, it did not initially prioritize training.

Christian Aid, as a smaller organization without field offices or a structure of advisory units at headquarters, from the start prioritized gender planning training of all staff with responsibility for project funding decisions. It saw this as a long-term process and, therefore, institutionalized it through three gender-aware generalists and a trainer. They all took on the responsibility for training as one of their mainstream activities. In addition, planning procedures changed because of redesigned checklists and the introduction of project-level gender guidelines. Both planning tools used the concepts and planning methodology introduced through training.

Both organizations say they have made progress in institutionalizing gender, but still have a long way to go. Both confirm that at the time the strategy adopted was the appropriate one for their particular organization. The strength of a WID unit such as GADU is that it can act both as a monitoring unit to ensure integration, as well as a pressure group to ensure that WID is constantly on the agenda. A recent survey of NGOs in Europe confirms these findings. It showed there was a correlation between the existence of a person, group or department dealing with gender issues and the formulation of WID policy documents (LCD 1989). In the same vein, proponents of separate organizations argue that the strategy to mainstream hides the fact that WID is not a sufficient priority to warrant a separate unit. Reliance on mainstreaming also can mean that WID has to challenge overriding concerns in each sectoral interest. It therefore is more likely to be institutionalized into areas in which women already predominate (such as health, education, welfare and nutrition) than the more technical sectors of the institution.

Anderson and Chen (1988) have argued that it is not possible to predict if effective programming will occur based solely on which institutional model an agency adopts. While it may not decide the outcome, the type of institutional model chosen reflects organizational 'culture'. The evidence suggests that a mainstream approach is more likely to occur in organizations with an organizational 'culture' of consensus management. The creation of a separate unit is far more likely to be advocated in organizations with a 'culture' of conflict management. The creation of a separate unit frequently results in higher levels of organizational tension and trauma than does a mainstream strategy. The holder of the WID post has a mandate to confront those not delivering, but also must be prepared to be on the receiving end of abuse. The staff of GADU are clear that the creation of a separate institution has been critical in providing the necessary space to think through WID policy. Without such a unit they could not have formulated policy.

However, they acknowledge that it has not been smooth riding. One staff member commented that it was a miserable job and could have been called the moan and groan unit. Staudt argues that 'Gender redistributive politics are as conflict laden as any other redistributive issue, but are subtle in the personalized resistance they incur, and complex in their confusion with cross-sex interpersonal relations' (1990: 4). Where the issue is identified in terms of an individual, generally a woman, the conflation of the professional and the personal is often far greater. Open displays of conflict may be more cathartic in changing attitudes, but are certainly more painful. A clear cycle of changing attitudes is visible in many organizations, and Oxfam is by no means unique in this. When gender is first raised as an issue it is often dismissed as a joke; then personal attacks occur that question the sexuality of individuals; finally, acceptance mainfests itself in a pretence of agreement. A member of GADU commented that 'When GADU started many people laughed; one or two called us lesbians, dykes; that is no longer acceptable. Now the most difficult problem is one of people pretending they agree.'

This pressure to personalize is widespread. Staudt provides an example in a case study of USAID. She cites testimony before Congress, in which a Representative repeatedly tried to prompt and extend the words 'male chauvinism' and 'male chauvinist pigs' to the WID co-ordinator's comments. In response, she replied, 'It's your term, not mine' (Staudt 1983: 270). Open displays of conflict are disturbing for both men and women. Men employed in WID units often experience a very particular form of trauma. Especially with their other male colleagues in the organization, they have somehow become de-sexed by the professional issue on which they are working. The evidence suggests that a mainstream approach does not seem to provoke such problems. Here the concern is to identify who has responsibility for WID. The danger is that issues will not be monitored because everyone simply passes the buck. As a staff member of Christian Aid commented, 'The fact that people pay lip service to the rhetoric of gender does not mean they necessarily take on the issues in practice. They are not necessarily fully on board.'

It is still important to clarify principles. Nevertheless, it has become clear that the amount of action on WID at the international level has lessened the relevance of the debate concerning alternative institutional strategies. The adoption of the Nairobi Forward Looking Strategies requires governments to set up separate 'machineries'. The OECD DAC Revised Guiding Principles on WID mandates administrative adjustments to help implement WID policies. Both require the identification of personnel with WID responsibility. This has meant that even in those organizations with a mainstream

policy, representation at international monitoring meetings requires a designated individual, by default if not design, to wear the designated WID hat.
At the NGO level similar trends have emerged. In Christian Aid the generalists in the gender planning training team have to resist very hard not to
become, *de facto*, the WID Unit. The lack of a person with full-time WID
responsibility has implications for Christian Aid's impact at the level of the
inter-NGO lobbying and networking. In some instances staff members
represent Christian Aid as a function of their job, while in other cases it is
the Women's Forum's responsibility to be involved.

If the issue is not the validity of WID units as such, then the critical
question is the determination of WID's power, legitimacy and mandate.
Obviously, there is no single factor but several causally interrelated constraints and opportunities that have effectively determined the capacity of
women's units to institutionalize WID. The following section discusses each
separately to show the manner in which each influences the other.

Is its structural location a problem? the experience of government women's bureaux

Anderson and Chen argue, 'it is not the institutional machinery *per se* that
makes the difference ... what is important in whether WID is taken seriously
is the institutional strategy that its proponents adopt' (1988: 4). If the creation
of a separate unit is not the problem, its location within the planning structure
is often the first important decision in the institutional strategy. In government women's bureaux, Gordan (1984) has argued that ministerial location
reflects the extent of the government's commitment to the notion of integrating women into national life. It also reflects its ideological position about the
role of women and, therefore, its perception of the role of the machinery itself.
Power and legitimacy relate primarily to the machinery's access to the
planning, policy formulation and funding sectors of the government. These
determine its ability to have any significant impact on the direction of
development in the country (Anderson and Chen 1988).

Without exception, women's 'machineries' of government were created
'top-down' by the political leadership as a response to the demand from the
international community during the Women's Decade. This meant that the
creation of a women's bureau was often more a symbolic gesture than a
legitimization of women's activities (Gordan 1984). Nevertheless, its structural location was the logical consequence of the current policy approach to
women by the government concerned. Since this was almost without exception welfarist in orientation, the common pattern was to locate women's
bureaux in ministries of social affairs or welfare. One exception, however,
was the location of nearly a fifth of all bureaux in the prime minister's office.

This was largely a matter of short-term political expediency. Only recently, with the shift towards an efficiency approach, have governments begun to consider the importance of women's units in such ministries as agriculture or industry.

The structural location in ministries of social affairs or welfare over the past decade has been highly problematic for two main reasons. First, there is a fundamental contradiction between the UN Decade's objectives for machineries to improve the status of women by ensuring their full integration into national life, and the welfarist objectives of social welfare ministries. Their concerns, in contrast, are to ameliorate the position of women as passive beneficiaries of development, while reinforcing their subordinate status. Secondly, because ministries of social welfare, as 'soft', marginal ministries, have traditionally lacked power, status and associated financial resources, these same problems have been transferred to women's bureaux.

Not only has structural location proved problematic for women's bureaux, but in some countries the constant moving of location has probably been even more so. In her 1984 study of six Caribbean countries Gordan identified that the three earliest countries to set up their women's machinery had each had four changes of location.[2] This might be expected in the early stages of institutionalization of WID, particularly with the high profile expected in 1985. However, Table 6.1, which updates, and extends, Gordan's study to include nine Caribbean countries, shows that this problem continues.[3] While the average for all nine countries was approximately three location changes, five of the nine countries had experienced four or more. This may be the consequence of political expediency, and reflects the continuing marginality of the women's bureaux. It also suggests that governments have experienced considerable problems in integrating the cross-sectoral issue of women's concerns into sectoral-based planning systems. The constraint can be identified as both political and technical in nature.

The Grenada government, for instance, has repeatedly redefined the sectoral basis of women's concerns. It moved its Department of Women's Affairs first from the Ministry of Education and Culture, to Health, then to Housing and Community Development. Communications and Works followed, and it has ended up in Tourism and Civil Aviation. Equally, Trinidad and Tobago has struggled to redefine its cross-sectoral ministerial concerns. It has moved its Women's Bureau from Social Security and Co-operatives, first to Health, Welfare and the Status of Women, and most recently to Social Development and Family Services. Those women's machineries with a lower level of mobility, such as in Belize and the Bahamas, have located them from the start in ministries broadly concerned with social development.

Constant changes in location can only be detrimental for women's bureaux. Not only have they involved physical movement and often loss of staff,

Table 6.1 Location of women's machinery in ten selected Caribbean countries (1974–89)

Country	Bahamas	Belize	Grenada	Guyana
Name of Machinery	♀ Affairs Unit	Dept. of ♀ Affairs	Dept. of ♀ Affairs	♀ Affairs Bureau
1974				
1975				
1976				
1977				
1978				
1979			Education & Culture	
1980				
1981	Youth Sport & Comm. Affairs	Social Services & CD	Health, Housing & CD	Labour
1982				Bureau of Cooperatives
1983			Communication & Works	Prime Minister's Office
1984				National Mobilization
1985			Tourism & Civil Aviation	
1986				
1987				
1988				
1989				Culture & Social Services

CD = Community Development
♀ = Women's

Table 6.1   (Continued)

Jamaica	St Kitts	St Lucia	St Vincent	Trinidad
Bureau of ♀ Affairs	Min. of Health and ♀ Affairs	♀ Affairs Division	Min. of ♀ Affairs	
Youth & CD				
Prime Minister's Office				
Health & Social Security				
Youth & CD				Social Services & Co-operatives
				Health, Welfare & Status of Women
	♀ Affairs		Tourism	
Labour, Welfare & Sports		Community Development, Social Affairs & Youth		CD & Status of Women
			Education	
	Health & ♀ Affairs			Social Dev. & Family Services

but also a constant redefinition of the programme to coincide with the interests of the new minister. Where the rationale for location is particularly obscure, as with tourism, youth, aviation or sport, there is not a coincidence of interests, but more a competition for scarce resources. However sympathetic the ministry concerned may be, moving also involves the disruption of the work programme while new bureaucrats need gender-aware training, using time and costly, scarce resources. Lack of delivery by the women's bureaux then simply serves to reinforce the organizational ideology about lack of professionalism and consequently the lack of importance of this issue.

UNCSDHA, among others, has argued that because the issue of location is obvious it has received more attention than it merits. The real issue is the more basic one of access to power, achieved by a variety of means, only one of which is location (UNCSDHA 1987). Despite their specific location, women's bureaux work within an operating bureaucratic system. In principle, this gives them the right of access to available resources of services, personnel and machinery within the system. Gordan (1984), therefore, concluded that political support for WID may not be as important for progress as appropriate strategies. Bureaux' success depends on an administrative framework that permits collaboration with the top bureaucrats and technocrats in all relevant ministries. One important measure has been the identification of administrative arrangements emphasizing multisectoral bureaux functions. The extent to which this has been successful has depended, above all, on the policy framework of the women's bureaux.

Is lack of coherent policy the problem?

As outlined in Chapter 4, different policy approaches to WID, each with specific goals, objectives and strategies, have developed over the past two decades. These have important organizational implications. Clearly, many organizations that created 'top-down' women's bureaux and units lacked a clear policy approach to WID. So did most WID units when first formally constituted. Their broad goals reflected the popular international rhetoric of the Women's Decade more than specific country-level priorities. The Decade's extensive goals of equity, peace and development all concurred in their aspiration to improve the 'overall status of women' (equity) and to 'integrate women into development' (development). However, no clear consensus existed as to the strategies by which this might be achieved. The 1985 Forward Looking Strategies provided more specific objectives, and were important starting points for government-level women's organizations. Nevertheless, it was the responsibility of each individual government to develop its appropriate strategies. To date, in many countries all that exists are broad policy statements. These lack specific goals and priorities, or the relevant planning procedures and management support structures necessary to transform them into practice.

Where goals, objectives and strategies have been defined, there is often conceptually a lack of internal consistency. For example, although endorsing the rhetoric of equity, few women's bureaux have understood, or acknowledged, its implications in terms of its top-down priority of women's strategic gender needs. In practice it results in goals of equity combined with strategies of welfare. The reasons are obvious. Welfare policy is still the common WID approach. Many governments, and the personnel recruited into the women's bureaux, do not acknowledge that welfarist measures help low-income women meet practical gender needs, but reinforce their subordinate gender position. These approaches are unlikely to reach the strategic gender needs agreed in the overall goals.

In the past decade organizations have struggled to define specific goals, objectives and related functions. For many bureaux, established because of government lip-service to the Decade, their high profile and the level of expectation to deliver on every front exacerbates the situation. A common tendency has been the simultaneous adoption of numerous strategies, with activities ranging from research on women, through policy formulation to project implementation. In reality, however, one institution cannot be expected to provide the full range of policies, programmes or projects necessary to accomplish WID goals.

Anderson and Chen (1988) argue that agencies differ in their capabilities and limitations. Each may have relative advantages over others in terms of the level at which it works. They distinguish between three types of institutions, government, professional intermediary institutions and private voluntary organizations (NGOs), and summarize the relative strength of each. NGOs are strong as advocacy and promotional agencies. They have organizing skills but fewer technical and managerial skills and, therefore, can play an important advocacy role in raising and popularizing issues. The strength of professional institutions is in their research and ability to provide technical and managerial assistance and, therefore, their importance is in providing assistance for targeted programming. Finally, governments have greater access to financial resources and specialized expertise. This gives them advantages in terms of scale and coverage and infrastructure development necessary for integrated, mainstream and broad-scale programming.

The 1990 survey of women's bureaux in nine Caribbean countries revealed that all implemented projects. The majority also undertook research and advised other ministries, and about half also undertook government policy formulation. Although in theory different organizations have relative advantages in terms of their mandate, in practice political legitimacy forces women's bureaux to undertake a wide diversity of strategies. A fundamental policy problem for women's bureaux is the lack of integration of WID policy into national policy objectives. Although paying lip-service to WID policy, governments in the main have continued to see women as an add-on or separate matter, purely the concern of the women's bureaux. The failure to establish women's priorities in terms of other national priorities has important institutional implications. It means that the women's bureau operates in a highly hostile environment in the ministry where it is located. Other departments assess it as an imposition and drain on resources, rather than as a contribution to the profile or status of their ministry. They therefore do not make the arrangements necessary to ensure collaboration between the bureau, its parent ministry and other governmental bodies. Without defined lines of responsibility and accountability, professional staff cannot operate. Gordan's (1984) assessment showed that difficulties experienced in getting regular contact with the Permanent Secretary of their particular ministry was one important problem identified.

The failure to integrate WID policy into national policy objectives results in a lack of clarity as to the role of other government agencies in carrying out policies relating to women's needs. Problems in persuading other bureaucracies of the fundamental need to link welfare concerns about women with the major economic, social and political concerns of their country are widespread. Finally, in the creation of intra-organizational linkages, so critical if women's bureaux are to move from project- to policy-level, WID units have experienced severe problems. These relate both to identifying administrative structures for intra-governmental linkages as well as those with relevant NGOs. Ministries often lack the necessary consultation processes to ensure that they incorporate women's concerns into their work. They often achieve more effective programme co-ordination when they institute routine administrative arrangements such as steering committees and working parties. However, these are only successful if officially validated with precise objectives specified. Even where such legitimacy is provided, there are enormous organizational problems associated in the co-ordination of cross-sectoral planning in ministerial structures based on sectoral planning procedures.

Is control over resources and limited staff the problem?

Since power and status within organizational structures is most frequently determined in terms of control over resources, this is commonly identified as the most important factor in institutionalizing WID. Within organizations different parts compete for budget allocations. Consequently, the size of the budget has come to exemplify the organization's perception of the importance of the work of the WID Unit. Many governments regard the budget statement as the annual declaration of its policy. This means that the only other important source of leverage for women's machineries is external funding from either official donors or NGOs.

There is a consensus that all women's machineries are severely under-funded, with participation in their organization's budgetary planning marginal, if present at all. The evidence is not only fragmentary but also difficult to evaluate. Budget allocations vary depending on how far the bureau has an implementing as against advisory function. The 1990 Caribbean survey, for instance, showed that all the women's machineries had an annual budget allocation. However, the amount involved varied very widely, in relation not only to its status, but also its mandate.

One consequence of limited budget allocation is that it has forced women's bureaux to rely increasingly on external funding. This in turn has meant less control over their preferred policy direction. The fact that some donors prefer to allocate resources to projects, particularly income-generating projects, forces bureaux to become more project-orientated than their mandate designates. This reduces their capacity to fulfill other objectives. It is important, however, not to over-exaggerate the size of resource allocation made by donors to women's projects. In the United Kingdom, ODA resource allocations for WID-specific projects increased from four in 1987/88 to six in 1988. For WID-integrated projects it increased from two in 1987/88 to sixteen in 1988 (ODA 1989). There are considerable problems caused by fixed short-term project funds, that leave bureaux to carry on without adequate support. External funding, however, is often critical for political leverage to obtain recognition and consequently funding from internal sources. Consequently, bureaux constantly make difficult decisions in terms of very different priorities.

As the mandates of women's units and bureaux shift more towards influencing policy than carrying out projects, their concern becomes less their own budget allocations, and more the WID allocations within other parts of the organization. In ministries of agriculture, for instance, what are the budget allocations for projects involving women heads of households? In ministries of health, how are resources allocated between large curative teaching hospitals and mother and child health centres? As the mandate of the women's unit becomes one of advising, monitoring and evaluating that the organization integrates gender in all work, their biggest budget expenditure then becomes staff salaries.

Budget allocations vary depending on the extent to which the women's unit is an implementing agency. Consequently, the best indicator of status and power is that of staff size. Despite the nature of the institution staff allocations to WID are universally pathetic. The 1990 survey of Caribbean countries reveals an average staff size of four (excluding Jamaica, with the unusually large staff size of fifteen). A recent OECD survey of WID officers employed full-time in thirteen counties with WID Units shows an average of 2.4 (excluding the United States with its atypical number of sixteen) (OECD 1990).

Other important issues need to be mentioned in relation to staffing: first, the role of the director. Much has been written of the deciding role of the director whose charismatic power or hopeless incompetence influences the unit's outcome. Clearly, an inordinate amount of responsibility for the potential success or failure of the women's unit is vested in the hands of one person. Her (or his) job description covers everything from academic knowledge of WID policy, through management skills to competence in publicity, funding and international relations. Yet the status and salary level of this job does not make it necessarily very attractive to many professional women unless they are ideologically committed to working on women's needs. In the 1990 Caribbean survey, the average age of the head or director of women's bureaux is thirty-seven years of age. Of these, 55 percent have higher educational diplomas or certificates (in social work, nursing or teaching), with the remaining 45 percent holding university bachelor degrees in the social sciences.

In many agencies, government and donors alike, women can become stigmatized and stuck if they work too long on WID issues. The labelling of women WID officers with their work is an example of the conflation of the personal with the professional that so dominates this career structure. In the case of donor agencies it is not the aid agencies that suffer, although they become duller places in which to work. It is women in developing countries who will lose badly needed allies in the donor community (Himmelstrand 1990: 111).

An indicator that also highlights the hypocrisy of the WID mandate of so many donor agencies is the small number of women employed on their staff. In her discussion of the bureaucrats who bring life to an institution, Staudt argues that there is sufficient evidence to be optimistic that more feminists inside bureaucracies can make a difference for women outside. This is particularly the case when they incorporate and combine their own agendas with other activist agendas (Staudt 1990). However, it is important not to assume that women by definition are feminists. There are as many examples of women who have not necessarily served women's interests, such as those described by Yudelman (1990) in the Inter-American Foundation, and Helzner and Shepard (1990) in Population Private Voluntary Agencies. Equally, the co-option of women's organizations into political parties, which really falls outside the discussion in this chapter, is both widespread and well known (Hirschmann 1990).

Despite the problems identified above, it is still salutary to identify not only the numbers of women in such organizations, but also the positions in which they are located. Examples from donor agencies in the United Kingdom illustrate this point. In the Commonwealth Secretariat, for example, although women comprise half the total staff, only a quarter are professionals, of which top management comprise 10 percent. As in the majority of organizations, women predominate at clerical and secretarial levels (Commonwealth Secretariat 1988a: 28). Of the consultant technical co-operation officers (TCOs) recruited by the UK ODA to advise governments and parastatal organizations in 1987/88, some 11 percent were women, increasing to 12 percent in 1988/89. Even an NGO with a very specific equal opportunity staffing policy such as Christian Aid has had problems. In 1983 the Women's Group put forward the recommendation that equal representation in area secretaries, board members and regional committee members should be achieved in five years. However, by 1987 only 35 percent of area secretaries were women, as were 32 percent of board members, and 35 percent of regional committee board members (Christian Aid 1987). Although these figures were below the target set, they are far higher than in any other UK NGO, except in NGOs targeted specifically at women such as Womankind and Opportunities for Women.

For donors, another important indicator of the responsiveness of national governments to WID is the number of women sent on overseas training. The Technical Cooperation Training Programme (TCTP) of the UK ODA, for instance, in 1987 had a total budget of £65 million sterling. It involved more than 110 countries and 11,500 students, of whom some 15 percent were women, with this number declining slightly in 1988 (Holden 1988; ODA 1989). Table 6.2 provides a detailed picture of the percentage of the budget given to women students among those countries allocated the largest numbers of TCTP awards. The Canadian International Development Agency (CIDA), even with a much more specific positive discrimination policy, has only been able to increase slightly its numbers of female sponsored trainees and students. In 1985, its global figure was 23 percent, and although it increased to 32 percent in 1986, in 1987 this was down again to 29 percent (CIDA 1988).

There are obviously several interrelated reasons for the appalling problems experienced with funding. Lack of adequate policy influences the level of resource allocation. Typically, all that exists are broad policy statements, without any clear definitions of goals and priorities. Associated arrangements for the provision of proper levels of resources and management support structures do not exist. Furthermore, the institutional location of WID Units influences resource allocations. Women's bureaux located within welfare ministries are not only last to be allocated money but also the first to be cut. Here the problem is less the status of the unit itself, and more that of its

Table 6.2   UK ODA TCTP awards to women, 1987–88 (percentages)

Country	Year	
	1987	*1988*
India	6.0	4.8
Kenya	16.5	16.5
Bangladesh	8.0	8.0
Malawi	17.1	18.8
Uganda	18.1	22.6
Pakistan	13.1	7.1
Tanzania	23.8	22.2
Zimbabwe	16.7	18.2
Zambia	19.1	15.3
China	26.4	15.9

Source: ODA 1989

ministry. Besides such administrative problems, there are more fundamental political and ideological reasons for the severe budgeting constraints so widely experienced. Budget allocations are clearly the ultimate verification of legitimacy, when lip-service must be converted to practice. Consequently, it is the biggest challenge and the most effective way in which male privilege blocks women, and reinforces their subordinate position. Well-financed units, provided with adequate supportive structures, can serve as crucial institutional structures to ensure positive discriminatory practices and oversee the integration of WID into the institutional structure. As such, they are consequently threatening to the existing institutional system of power and control that effectively marginalizes women's gender needs, both practical and strategic.

Is lack of awareness of gender planning the problem?

Can you shift the ideology of the organization or the work practices of individuals through training? Recently, as more technical constraints in the institutionalization of WID have sorted themselves out, attention has increasingly turned to the more entrenched political constraints. Gender-blindness of both male and female colleagues working within such organizations, has increasingly been identified as one important problem. Training in gender analysis, gender-awareness or gender planning has become an important solution, particularly in those organizations that have adopted a mainstream approach. Since this is such an important area of current debate, the issue is raised here, with a detailed discussion of the debate around gender training outlined in Chapter 8.

Table 6.3 Alternative strategies for the institutionalization of WID/GAD

Type of strategies	Specialization	Combined strategies	Mainstreaming
Structural location	Specifically designated WID unit	WID unit works in collaboration with existing organizational structure to promote WID issues	Institutionalization within existing organizational structure
Targeted personnel	Gender specialists controlling WID mandate & providing expertise to others	Gender specialist acts as a catalyst throughout the organization Designated WID specialists monitor each mainstream office	Gender-aware generalists integrating WID/GAD perspectives into work
Priority interactions	Critical mass of gender specialists to monitor/watchdog the issue	Providing the support structure for WID personnel in the mainstream Incremental training in gender awareness/planning Development of gender-specific tools as well as gendering of existing operational procedures	Extensive training in gender awareness/planning Gendered changes in planning procedures and development of new methodological tools
Organizational 'culture'	Conflict?	Interactive?	Consensus?

INSTITUTIONALIZATION AS A PROCESS RATHER THAN A PRODUCT: OPPORTUNITIES IN CURRENT STRATEGIES

If it was politically naïve of women to think that the creation of a new institutional structure would ensure the institutionalization of a new planning concern, it was also politically expedient for planners to believe that these palliative measures would be sufficient to keep women quiet. A number of different stages in the process of institutionalization can be identified. The 1970s was the 'introduction' stage, when the WID issue was first legitimized within government bureaucracies with WID structures defined. This was followed in the early 1980s by a second 'fashionable' stage, during which a few women were appointed to small, under-financed units. Since the end of the Women's Decade in 1985 a third, 'co-option' stage followed, with the recognition that WID was not a short-term trend (Himmelstrand 1990). Organizations began to define their demands so that they fitted into established structures without requiring organizational changes. WID units were integrated into the co-opting establishment, and accepted by them as a regular, if peripherally important, member.

This chapter has examined the institutionalization of WID in a diversity of institutions in government, donor agencies and NGOs. The experience in each is obviously unique, with different constraints determining success. Nevertheless, most institutions are now involved in the third, co-option phase, with 'gendered bureaucratic subversion' the most problematic issue. A decade of experience has shown that the 'top-down', product-orientated creation of the WID Unit has not institutionalized WID. Most frequently, it has simply been co-opted into the dominant organizational structure. In the past few years, therefore, organizational structures have focused attention on the development of collaborative, bottom-up processes within their structures. These are attempting to identify blockages, and to use combined strategies. These include the creation of new, more effective organizational structures and alliances. Women's machineries have been forced to understand more about the nature and content of bureaucratic politics, while focusing less on the performance of their own WID Unit.

Combined strategies as an institutional strategy: the case of Swedish SIDA

By 1990 a convergence of trends had occurred in most agencies. The institutionalization of WID occurs through a strategy that combines the creation of a WID unit with a mainstream approach. A successful strategy generally has two main components. First, mainstream departments, field offices and other important parts of the organization appoint designated WID specialists. Secondly, the role of the WID Unit gender specialists changes. They give less emphasis to controlling the WID mandate, and more emphasis to creating the necessary networks and alliances with gender-aware sectoral specialists. Through the development of effective coalitions and lobbying pressure they give up their WID mandate. The most critical institutional issue then becomes the identification of gender entry points, with the best tactic to operate on several fronts simultaneously. Combined strategies are not necessarily new; what is, is the precise level of definition.

Again problems can easily occur, with combined strategies not necessarily successful. WID specialists can be as marginalized in mainstream departments as they were in the WID Unit if they lack the necessary collaborative support and networks. The long-term WID Unit gender specialists often have problems letting go of their ownership of WID, which has so long provided them with the basis of their legitimacy.

The experience of a bilateral donor such as SIDA provides a useful example of the *process* that has to be undergone in order to achieve a combined strategy. The WID Unit was established in 1979. Like Oxfam's GADU, it was a temporary arrangement to exist only until all SIDA's programmes under the supervision of programme officers had integrated the 'women's dimension'. Commenting on this, a recent head of the WID Office said, 'This was the watchdog period.' In the past few years its status has been confirmed as permanent, due to pressure, from both inside and outside the organization, that 'WID is not a trend of short duration' (Himmelstrand 1990: 110). It considers its location within the Planning Secretariat a good position since it has direct access to the Director General. Nevertheless, the Gender Office (as it is now called) has only five full-time appointments in Stockholm. Identifying themselves as *catalysts*, their institutional strategy comprises three clearly identified, interrelated priorities. First, they have focused on building up a support structure for WID personnel in the country-level Development Co-operation Offices (DCOs). The role of country-level officers is to ensure that sector programmes integrate WID, as well as administering direct support. In 1990 there were twenty field-level personnel responsible for WID. They were mainly local, part-time appointments with rapid turnover (many of them Swedish spouses of Swedish overseas staff). The problems with junior-level appointments are not solely that officers lack knowledge of SIDA's structure and procedures; of greater importance is the fact that their low status impedes them when trying to challenge more senior staff on controversial gender issues. To counteract these tendencies, the Gender Office would like to have two positions at programme officer level. This would consist of a Gender Officer responsible for integration aspects, until all SIDA staff are sufficiently gender-aware, and a Gender Adviser who ideally should be a national, responsible for work on direct support and acting as a long-term adviser on all programmes.

The second component is training in gender-awareness and gender planning methodology based on SIDA's case studies and related to SIDA's planning cycle. Heads of Sector Bureaux, Heads of Sections in Sector Bureaux and Heads of DCOs, as well as Programme Officers and consultants all undertake such training. The purpose is not only to make them gender-aware. It is also to introduce a planning methodology to them, which includes identifying the institutionalization of the WID officer in the mission's work. WID officers also undertake gender training. This helps professionalize them in their job, as well as assists them to identify critical entry points for WID strategy in the SIDA planning cycle.[4]

By way of comparison with SIDA's strategy, the OECD in its Monitoring Report assesses the creation of full-time positions for WID experts in the field to be one major achievement accomplished. Five countries currently appoint full- or part-time positions (ranging from the United States with seventy part-time positions to Denmark with four full-time). However, while the report states that there has been a noteworthy expansion in the countries undertaking this, the number of personnel is the same as two years previously. NGOs also comment that where a WID field officer has been appointed the likelihood is that gender issues will be more effectively integrated into projects, and more resources allocated to women's projects. While this is obviously a critical move to mainstream, clearly the importance of WID field officers relates directly to their status, and legitimacy both from the local field office as much as from headquarters. If their mandate is simply collecting information on women, this does not necessarily in any way impact on major policy decisions or programme- and project-level resource allocations.

The third component in SIDA's strategy is the development of methodological tools to be used in the different stages of SIDA's planning cycle (SIDA 1989). The Gender Office has its own small budget for supporting initiatives at national, regional and international level. However, inputs made within sector programmes are normally financed within the existing country frame. This means that budgetary issues therefore have become mainstreamed.

Challenging blockages and creating alliances to provide more effective political leverage

Combined strategies, such as those used by SIDA, clearly reduce many technical constraints experienced by those adopting single strategies. Indeed, combined strategies have been adopted widely, especially by multilaterals such as FAO and bilaterals such as Canadian CIDA and USAID. These latter two, in particular, now have comprehensive, widespread institutional strategies that also include the three components described above. Nevertheless, they are still essentially technical solutions and in themselves are often necessary but insufficient for fundamental changes in institutionalizing WID. Other measures are required if more fundamental political constraints are to be challenged.

Increasing the focus on intra-organizational networking

The importance of intra-organizational networking has long been recognized. Such alliances are necessary to create leverage in bureaucratic politics. However, networks are only useful if structured to contribute to mainstream institutional procedures. Many intra-organizational WID networks of governments, donor agencies and NGOs alike, were set up simply to legitimize WID, and are considered time-wasting and irrelevant by those involved.

In addition, different types of networks are simultaneously required. For instance, a recent monitoring report of the OECD DAC/WID group identified three different institutional structures, set up as support to WID units. Four OECD member countries have established high-level WID steering committees. These are hierarchically structured and composed of senior policy and operations departments staff entrusted with the responsibility for overseeing the implementation of WID policy, in both an advisory and monitoring role. Their purpose is to show a strong and visible commitment by the senior management to integrate gender needs and interests in bilateral concerns (OECD 1990).

Secondly, nine countries have set up Inter-departmental Task Forces on WID. These are horizontally composed of relevant policy and operations department WID liaison officers, with responsibility for different regional or sectoral divisions. Their purpose is to ensure the integration of WID policy into practice through the entire spectrum of aid activities. Finally, three countries have established *ad hoc* Working Groups, to work on specific WID-related issues as needs arise.

The location of WID Units has important implications for intra-organizational alliances. Set up as an advisory unit, Oxfam's GADU, for instance, has no mandatory power and influences procedures and policies through networking. The Unit initially assessed the blockages they encountered to be the consequence of the threatening nature of WID. Over time, however, they witnessed other advisory units on such non-threatening issues as health endure a similar fate. They have recognized that some constraints, although exacerbated in gender issues, relate to the structural problems inherent in the status of a Unit.

More effective use of the leverage of inter-organizational structures

The last decade has witnessed the growth and consolidation of inter-organizational networks and the development of a wider strategy to ensure effective organizational leverage. These work at different levels, involving different combinations of the three organizational groups discussed in this chapter, donor agencies, national governments and NGOs. Many were set up or reconstituted to provide additional leverage for implementation of the 1985 UN Forward Looking Strategies.

The OECD/DAC WID Expert Group has been an important catalyst for donor agencies, in raising the awareness of WID issues and the Forward Looking Strategies among member countries. Adoption by the DAC of the 1983 Guiding Principles on Women and Development (and the 1989 Revised Principles), as well as annual monitoring of progress, both provide important means by which pressure is quite clearly exerted on 'reluctant' governments, with particularly recalcitrant members invited to join the steering committee. The fact that nearly all member states now have WID mandates, have adjusted administrative and personnel arrangements to 'ensure the integration of WID concerns into their AID programmes', and have elaborated comprehensive Plans of Action has not necessarily resulted in fundamental changes. The 1990 Third Monitoring Report notes that the implementation of the measures stipulated in WID action plans still depends on the interest, understanding and willingness of individual staff members, with the changing of attitudes of some staff members still 'a slow process'.

A similar inter-organizational network to assist the implementation of the Forward Looking Strategies is that of Commonwealth governments. The Commonwealth Plan of Action accepted in 1987 by all Commonwealth governments contains guidelines for national governments as well as national plans of action for WID bureaux. The Women's Development Programme (WDP) at Commonwealth Secretariat headquarters in London monitors progress through an annual questionnaire. In addition, the WDP provides a funded extension service in response to requests for assistance from governments formulating their WID policy and Plans of Action. The Commonwealth Secretariat has a certain ability to pressurize national governments. However, compliance is still dependent on the political will of national governments, many of which do not respond effectively.

For many organizations it is the interrelationship between the different types of inter-organizational pressure that effects changes. In the late 1980s the UK ODA Social Advisers were able to use the DAC principles as the leverage to enable the then Minister of Overseas Development to maintain momentum in his 'top-down' pressure to move the debate within a reluctant ministry. However, the simultaneous 'bottom-up' lobbying of the UK women's NGO group, the National Association of Women's Organizations, provided the final impetus for carrying out important changes to institutionalize WID within ODA. Just as lobbying of senators was important for the legitimization of USAID WID, in the same way the lobbying of Members of Parliament with questions in the House of Commons has put pressure on the ODA to shift its approach to WID (Commonwealth Secretariat 1988b). Similarly, Swedish SIDA acknowledges the critical bottom-up pressure provided by the NGO Council of Swedish Women for WID, whose monitoring of SIDA's progress with their WID agenda provides an important form of leverage. Finally, inter-organizational groups are also important between First and Third Worlds. As a staff member of a Northern NGO commented, 'When a woman from a Palestinian NGO says "Palestinian women are concerned about feminism" it's very helpful for us.'

Confronting blockages in bureaucratic resistance

Widespread criticism of women's bureaux, organizations and units have concerned their lack of policy goals, objectives or strategies. In the past decade, an important priority for most agencies, therefore, has been the clarification of policy. Accompanying this has been the development of timetable objectives, and specific proposals for action (to be analyzed methodologically in Chapter 7). It is also important, however, to identify the extent to which increased professionalism of policy and planning has had an impact on the institutionalization of WID.

The professionalization of policy

Many professionals working on WID are clear that clarification of policy has assisted them considerably in challenging blockages on a sounder professional footing. For instance, clear articulation of the different policy approaches to WID has enabled them much more effectively to identify and confront colleagues using inappropriate WID policies, particularly welfare. It has also allowed professionals themselves to develop a much wider strategy in the policies they endorse. It is here that the difference between WID and GAD has significant political implications. Some agencies, such as USAID and the World Bank, still tend to use a WID framework, focusing on women as separate target groups, and identifying efficiency as synonymous with equity. Others, such as SIDA and ODA, have worked to adopt what they define as a 'gendered' efficiency approach, while also recognizing the different tactical advantages of both the empowerment and equity approach. At the other end of the spectrum are NGOs, some of whom perceive a priori that all their interventions empower their target groups. They have rudely woken up to the fact that this does not necessarily include women. Obviously, the culture of the bureaucracy influences the tactics adopted, with concern focusing on the relative advantages of flexible, as against highly structured, institutions. While an *ad hoc* policy means that individuals have a greater scope to create new systems, they also are less accountable and have a greater capacity to block policies. In one Northern NGO, the WID officer named three individuals (all men) whose autonomous control over important areas of the institution, such as research, communications and evaluation, meant that they effectively blocked critical areas for WID policy. The problem with much of the professionalization of policy is that it depends above all on the skills of individual professionals. A very thin borderline separates playing the system with unstated agendas, co-option or blockage.

Confronting the resistance of men and women in the bureaucracies

Ultimately the professionalization of policy raises the impossibly sensitive issue of how to confront male bureaucratic blockages. Staudt has commented, 'For redistribution to occur, advocates will probably need to ally with men who are either unthreatened by power sharing or supportive for ideological, professional or technical reasons' (1990: 10). Certainly, the anecdotal experience of women in different agencies is full of examples of the critical, circumstantial as against structural leverage that a sympathetic man in a gate-keeping position can provide. There are, however, severe problems with the patronage that this type of leverage creates. The reverse is also true of women within the organization. Not all of them necessarily wish to represent women's interests. This provides yet another example of the conflation of the professional and the personal. In one UK NGO, for example, the intra-organizational women's group has the official status necessary to challenge critically and to monitor the organization's policy towards gender issues. However, the fact that most senior women in the organization chose not to be members considerably reduces the power of this group. This is a classic example of contradictions in the composition of intra-organizational groups, which concern not only status, but also the sex of participants.

One serious problem faced by donors and Northern NGOs alike is the fact that male staff still identify feminism as a Western concept. They themselves often have little idea of the prevailing attitudes of women in the Third World countries in which they work. A senior adviser in one donor agency has introduced the concept of 'environmental scanning' to her male colleagues. This is a polite method to invite them to 'test the water' and identify with greater rigour the status and position of women in the countries where they allocate aid. It would appear that middle management is generally the biggest problem, least willing to take the plunge and most likely to export their own role models into the contexts in which they are working. This is one area in which training has played an important part in changing attitudes.

The bottom line: incentives or sanctions

Those whose experience has taught them that change only comes out of self-interest dismiss the long list of issues discussed above as of limited impact, or useless in institutionalizing WID. For them the only recourse is to introduce incentives to encourage new attitudes, or sanctions to prevent the worst of bad practices.

Staudt (1983) argues that implementation occurs in response to appropriate incentives. The strength of positive incentives ranges on a continuum from prescription, to resource provision alone, to resource provision tied to performance. Negative incentives, such as sanctions in the form of veto capability or funding termination are potentially strong in molding behaviour. However, they often incur ill will and new, more subtle forms of resistance. Moreover, they are vulnerable to political pressure and rendered meaningless if applied inconsistently or infrequently. There has been an extreme reluctance in organizations directly to introduce incentives or sanctions to encourage new attitudes. One agency bravely embarking on this sensitive path is Canadian CIDA. It has introduced WID as one of nine criteria on which supervisors and managers are evaluated in their performance review and appraisal report. Since management responsibilities include 'ensuring that policies relating to Official Languages, Employment Equity, Performance Review and Employee Appraisal are followed with their area of responsibility', supervisors and managers are rated from one to nine on their 'Employment Equity' – namely, action on policies relating to under-represented groups, such as women, aboriginal peoples, visible minorities and the disabled. It is still too early to measure the effect, if any, of this on changing resistance within bureaucracies.

Concluding comment: the bureaucratic structure as the ultimate problem?

Ultimately, however effectively an institution has become gender-aware, or gendered, its institutional structure may still not have changed. Gender has simply been grafted onto the existing structure. This raises fundamental questions, still largely unresolved, concerning the extent to which engendering an institution ultimately requires a fundamental transformation in the structure itself.

As Cardon has noted, the principle by which women's groups operate is 'one of the few areas in which women's liberation has specified the application of its ideology' (1974: 86). This includes principles such as self-realization, equality or anti-elitism, sisterhood and authority of personal experience. All have encouraged structureless groups, 'uninhibited' by set agendas. The superiority of groups such as these lies in their being a non-competitive, supportive forum conducive to confidence-building and equal participation (Button 1984). This is in direct contrast to bureaucracy, the most rational form of organization through which legitimate authority is exercised. In its ideal form, Blau (1956) characterizes it as a hierarchy of authority, a system of rules, impersonality, employment based on technical qualifications and constituting a career, and efficiency. Feminists argue that hierarchies are antithetical to feminism. However, the degree to which women's organizations and hierarchies can co-exist is not proven. At the end of the day, some level of hierarchy, with accountability and effectiveness, is necessary. To date, the non-hierarchical structures of feminist institutions have made few inroads into existing structures, nor sufficiently proved their professional competence to deliver and implement in practice.[5]

Women's organizations recognize the need to talk to bureaucracies to keep their interests and needs on the political agenda. To do so, organizational forms acceptable to both groups, which can articulate their demands, need to be developed. Freeman (1974) has argued that while the leaderless, structureless group was part of the evolution of the women's movement, as the goals change from consciousness-raising to social action so also must the organizational form change. However, to suggest that women should be prepared to formalize their organizations is not to agree that traditional forms are superior. There must be other frameworks for building organizations based on democratic structuring and political effectiveness.[6]

Ultimately, if existing bureaucracies, as arms of the state, cannot or will not empower women, no matter the extent to which they are gender-aware, it may become necessary for women to empower themselves through very different forms of organization. This issue will be further examined in Chapter 9, the conclusion, which looks in greater detail at the constraints and opportunities of different types of women's organizations.

7 Operational procedures for implementing gender policies, programmes and projects

Operational procedures designed to translate planning into practice are a simultaneous concern, along with the creation of institutional structures for a new planning agenda. Can gender simply be 'added on' to existing planning procedures such that ongoing policies, programmes and projects are made gender-aware? Or are separate planning procedures required, along with specific gender policies, programmes and projects? Or is a third alternative the transformation of existing planning processes, through 'mainstreaming'?

Successful gender planning depends not only on an appropriate problem analysis and policy formulation but also on an adequate methodology to implement the policy. Over the past decade professionals working on WID issues have clearly articulated the constraints and opportunities in institutionalizing gender planning within existing organizational structures. They have fought for increased resources and new institutional structures, as described in Chapter 6. In marked contrast, far less attention has focused on the operationalization of gender into current planning practices. Traditionally, operational procedures in the planning process have been identified as the 'technical' domain of planning, in which the major problem identified has been a lack of adequate procedures. The response by those involved in WID issues over the past five years has been to ensure that WID policies, programmes and projects have been developed, along with associated planning tools such as guidelines, manuals and checklists.

It is only recently that gender planners have realized that the problem no longer relates principally to a lack of gender policy, or to partially formulated gender policy. The rigour of current gender diagnosis and analysis has helped to ensure that 'technically' such policy is becoming increasingly sophisticated. The most important problem now is the inability to translate gender policy into practice. Why does this so frequently fail to occur? Why does the content of WID/GAD programmes and projects so often change during the implementation process? These are the questions that need to be addressed.

Chapter 5 outlined the evolution of different planning traditions and their associated planning methodologies. This highlighted the difference between two planning traditions: first, rational comprehensive planning, with problem-solving technologies based on rational procedures and methods for decision-making; and second, 'transformative' traditions, that integrate conflict and negotiation into the planning process. In reviewing current practice it is necessary to identify the extent to which constraints in operational procedures are technical in nature, relating to inappropriate planning procedures. Is the problem that inappropriate rational comprehensive methodologies are being used to operationalize gender concerns, or are there wider political constraints which impede successful implementation? Or are both technical and political constraints and opportunities themselves linked in an iterative process?

During the past decade a range of organizations and agencies, at international, national and NGO level, have introduced new procedures into their planning processes, to operationalize gender concerns. From the wealth of experience of the past decade this chapter discusses different levels of intervention. Analysis of two very different planning interventions follows, with examples from both UN planning procedures, as well as from OECD bilateral agencies concerned with Third World development issues. The analysis concentrates primarily on constraints in operational procedures, with a detailed examination of the introduction of gender into the project planning cycle. Given the importance of project planning in the donor community, many see this as one of the most important areas of intervention. The examples come from personal experience and data accessibility. However, they also reflect the fact that several donors, both in Europe and in North America, have allocated considerable resources to the development of sophisticated tools and techniques. Assessment of their progress is intended to help other agencies concerned with gender planning within their own organizations.

HISTORICAL BACKGROUND: THE DEVELOPMENT OF WID POLICY

Since colonial and post-colonial governments first identified the problems of Third World, low-income women as a policy concern, many types of interventions have been designed to help them. However, until recently there has been very little systematic classification of these various policy initiatives. Chapter 4 examined the conceptual rationale underlying different policy approaches from a gender planning perspective. It categorized five WID policy approaches – on a continuum from welfare, through equity, anti-poverty, and efficiency to empowerment – and evaluated each in terms of the roles recognized and gender needs met.

The operational procedures and levels of intervention associated with each policy approach are not necessarily similar. For example, the early welfare approach did not articulate itself as WID policy as such. Ministries of welfare or charitable groups made interventions at programme or project level. They targeted 'vulnerable' groups that included women, rather than specifically targeting interventions to women. The first defined WID policy, at donor level, was the 1973 Percy Amendment to the US Foreign Assistance Act. The history behind this equity-approach initiative and the widespread antagonism it encountered was described in Chapter 4. So was its particular focus on top-down statutory intervention at policy level to improve the position of women through legislation.

Despite the apparent limitation of equity as a policy approach, its importance lies in the fact that it was the first approach to highlight the need for the identification of policy direction through a formulated WID policy. As described in Chapter 5, this is the first stage in any gender planning process. It is now broadly recognized that a clear, precise and unambiguous policy-level statement of the importance of WID provides the starting point for operationalizing WID concerns. At the same time it is important to realize that the particular policy approach to WID adopted, and consequently the content of such policy, may differ widely.

From plans of action to forward looking strategies: the experience of the UN system

Following on the formulation of WID policy has been the development of plans of actions and sets of guiding principles. The purpose of these has been to define more specific goals, priorities and strategies for the integration of WID issues into particular policies, programmes and projects. Several planning initiatives by the United Nations, linked to both the 1975 International Women's Year, and then the 1976–85 UN Decade for Women, illustrate these changes. They reflect not only increasing levels of professionalism, but also changes in planning methodology.

The UN World Plan of Action for the Implementation of the Objectives of the International Women's Year (UN 1976a) was an early initiative intended to provide guidelines to national governments for action over the ten-year period up to 1985. It was the General Assembly that proclaimed this the Decade for Women, with its objectives equality, development and peace.[1] Produced at a time of widespread legitimacy of national plans, it is a classic example of the blueprint 'survey–analysis–planning' plan to introduce women's issues into the political agenda. The methodology comprised three clear stages: of survey collection of data on women, its analysis, and the provision of recommended solutions. The Plan of Action outlined proposals for general national policy, as well as identifying nine key areas for action, with recommendations addressed primarily to government. For example, in the education sector the survey revealed that 'in many countries women and girls are at a marked disadvantage'. The analysis showed that 'this constitutes a serious initial handicap for them as individuals and for their future position in society'. The action proposal included the provision of 'equal opportunities for both sexes at all levels of education and training according to national needs and international standards' (UN 1976a: 15–16).

In each sector the Action Proposals provided an eclectic checklist of measures without setting gender objectives or indicating the entry strategy for their implementation. Some measures advocated were for such practical gender needs as adequate training or child-care facilities. Others, intended to reach controversial strategic gender needs such as equality of opportunities in education and equal pay rights, were unlikely to be implemented in practice.

The UN Plan of Action provided a model for many governments to develop their own national policy statements and plans of action. Jamaica, for example, was one of a number of Caribbean countries that extended its national policy statements to include not only general principles but also immediate goals and proposed measures. In each case it identified the responsible agency. Thus, in the Jamaican Bureau of Women's Affairs National Plan of Action, goal number 5 states:

> recognising that the evidence of physical and sexual abuse within families and societies is increasing – the government will pursue means of providing adequate protection and means of redress to women and children who are victims of family violence, incest, rape and sexual harassment.
>
> (Government of Jamaica 1987: 17)

It then identified several important constraints to achieving this goal, including the fact that social attitudes condone physical abuse within the family. The proposed measures included the implementation of an act for the prevention of abuse to women and children to be administered in the family court. The Ministry of Labour was identified as the agency responsible for this measure (Government of Jamaica 1987). However, further stages in the implementation process were not outlined, which raises the question whether these measures were seen simply as procedural in nature.

Important though such policy initiatives have been, in practice it has proved very difficult to implement many recommendations. One negative outcome has been that this has often served simply to reinforce the prevailing cynicism as to the relevance of gender issues to mainstream national planning. This manifests itself in hostility of the opposition, and the despair of women's organizations, who have lobbied long and hard to get such issues onto the agenda. To what extent is the planning methodology responsible for this problem?

In their evaluation of the Women's Decade, Tinker and Jaquette commented that it not only 'promoted and legitimised the international women's movement'; it also 'required attendance by governments at the three world conferences [and thereby] elevated women's issues to the level of international diplomacy' (1987: 419). A third very important effect was its influence at the international level on planning procedures to operationalize gender concerns. The 1985 Nairobi conference recognized the limited impact of the Plan of Action during the decade in terms of its goals of equality, development and peace. Consequently the Forward Looking Strategies (FLS), adopted by 157 nations at the conclusion of the Nairobi conference, provided a far more specific set of guidelines and recommended strategies for adoption by UN member states. It was produced as a result of comprehensive surveys and questionnaires completed by all member states on progress and obstacles to women's development. It outlines concrete proposals and actions to be followed between 1985 and 2000 and makes impressive reading, containing as it does a substantial agenda relating not only to practical, but also to strategic gender needs.

The FLS alludes to the essentially political nature of the planning process. However, as a UN document it is limited in its ability to set specific goals and targets, or to define entry strategies. It therefore, confines itself to putting into place a few international structures to monitor progress, and thereby indirectly exerting pressure on national governments. This it does through the Commission on the Status of Women. The focus for WID in the UN system is based at the Division for the Advancement of Women in Vienna. It monitors the implementation of the FLS to the year 2000 by preparing annual reports on WID progress for the Economic and Social Council of the UN (UNCSDHA 1989).

Finally, at the UN level it is necessary to mention the 'Convention on the Elimination of all forms of Discrimination Against Women'. This is the most comprehensive international legal instrument dealing with the rights of women, now ratified by some ninety member states. Unlike other items agreed by consensus at the UN assembly, ratifying this document means that a country has a legally binding commitment to work for the elimination of discrimination against women (Commonwealth Secretariat 1988b). The Convention requires that governments set up national legislation banning discrimination. Some thirty articles outline the international principles to implement equality in areas relating to political rights, public life, education, employment and pay, maternity rights, social services and child-care provision. The Committee for the Elimination of Discrimination against Women (CEDAW) supervises the implementation of the Convention. This is undertaken through monitoring and reporting procedures, requiring ratifying nations to provide reports at stated intervals.

The reporting procedures of both the FLS and the Convention have resulted in sophisticated gender diagnosis on a wide range of issues relating to the status of women. This may be their greatest achievement. Diagnosis is undertaken on a four-year cycle and, consequently, has the potential for integration into national planning cycles. Since national women's bureaux have responsibility for the data collection, this provides the opportunity for closer liaison and collaboration with other ministries.

Despite such measures, however, few countries have integrated gender into national development policy or planning procedures. In both the FLS and the Convention, government's participation in monitoring procedures is voluntary. The evidence to date shows that many fail to respond promptly, if at all, while many countries have yet to ratify the Convention. As INSTRAW (1988) has commented, it is only when women's organizations and others begin to monitor the work of CEDAW that the Convention will become a powerful tool for eliminating discrimination. Even in countries working to achieve absolute equality in the law, there is a wide gap between *de jure* equality and the actual state of affairs (UNCSDHA 1989).

From WID mandates to project-cycle handbooks: the experience of the OECD community[2]

The UN experience illustrates changes in planning procedures at the policy level, particularly during the 1975–85 period. This is applicable to an international institution with a mandate to monitor progress globally. Complementary to this is the experience of the donor community, discussed here in relation to the OECD countries. Particularly during the 1980–90 period, these countries had an impressive record in the development of complex planning procedures, and detailed tools to ensure that different stages in the planning process operationalize gender concerns. In describing these developments in Table 7.1, the purpose is also to clarify terminology and categories used.

The first critical stage in the formulation of WID policy is when organizations recognize the limitations of their gender-blind procedures and establish a *WID Mandate or Policy*. All nineteen members of the OECD's Development Assistance Committee (DAC) comply with the requirement that WID policies should be explicitly based on specific mandates. All therefore have such a document. This is a legally binding policy document, adopted at a high level, to show political commitment. Within its general guidelines, different options follow. These range from parliamentary law (Italy and the United States), to ministerial directives, to internal guidelines

Table 7.1 Planning procedures to operationalize WID

General procedures	Purpose	Type of country-specific procedure
WID mandate or policy	to define and legitimize WID policy	– parliament legislation – ministerial directive – internal guidelines – operational objectives
Plan of action	to provide WID operational strategies and procedures	– detailed plan of action – general action programmes – WID strategy paper
WID specific sector guidelines	to provide sector-specific WID guidelines	– sector-level WID checklists – WID sector papers – WID manuals
Integrated WID criteria in sector guidelines	to integrate WID criteria into general sector guidelines	– sector guidelines – WID in office procedure
Country-level WID guidelines	to integrate WID concerns into country-level operations	– country-specific WID plans – country plan of action – country development strategy statement – country programme/reports – country assessment WID papers
WID project guidelines	to integrate WID into the project cycle	– WID project cycle manual – WID project checklist – project identification document – project paper – checklist for participation of WID projects
Monitoring procedures	to monitor implementation of WID plan of action	– progress reports – built-in monitoring procedures – reporting to congress

emanating from the ministry or agency responsible for development co-operation (adopted by all other OECD countries). The scope of mandates varies considerably. Canada, for example, has one of the more complex mandates. An extensive policy framework defines the scope of the WID mission assigned to CIDA, as well as the orientation, operational objectives and strategy.

Within such mandates different approaches to WID reflect complex internal political processes undertaken to ensure that such a mandate is adopted. It also reflects the fact that policies themselves often are redefined over time.[3] Within the OECD countries the full spectrum of WID approaches exists. While most countries now have adopted a combined welfare/efficiency approach, others have what is termed a 'gendered efficiency' approach. Occasionally, when women are identified as a central target of development assistance, empowerment is more explicitly articulated.[4] For example, USAID, in a 1982 policy paper, clearly reflected an efficiency approach when stating that 'the key issue underlying the women in development concept is ultimately an economic one: misunderstanding of gender difference, leading to inadequate planning and designing of projects, results in diminished returns on investment' (1982: 1).

In contrast, a recent policy definition of the Netherlands Ministry of Foreign Affairs combines an emancipatory/efficiency approach:

> The advancement of women is a necessary component of development cooperation. It is required both by a fair approach to the female half of the target group, which has long been overlooked, and by an expedient approach to their substantial contribution, which has too long been under-estimated. Thus while the improvement of the position and status of women is fully valid as an emancipatory end in itself, the utilization of women's potential is at the same time an efficient means to improve the quality of development as a whole.
>
> (1989: 2)[5]

Recognizing that a WID mandate is not enough, OECD countries have sought to operationalize it through several different types of interventions (see Table 7.1). While mandates provide general guidelines, the next important procedure is a *Plan of Action*. Its purpose is to provide operational strategies to implement the WID policies outlined in the mandate. The plan identifies objectives and responsibilities over the full range of the agency's area of operations, and provides specific guidelines for the implementation of different departments and divisions. By 1990 all but two of the DAC members had adopted such Action Plans.

A further procedure is that of *WID specific sector guidelines*. These provide more stringent measures to ensure that WID is operationalized at the sector level. To date, most specific guidelines have been elaborated in areas where women's roles are identified as critical. These include agriculture, health, water and the environment. The guidelines highlight substantive issues at the sectoral level, also providing summarized checklists to help staff in drawing up terms of reference. Sometimes general action programmes or WID manuals include sector-specific details. An alternative procedure has been to *integrate WID criteria into general sector guidelines*. This procedure obviously involves greater collaboration and negotiation between WID and other sector experts. Again it is more likely to be undertaken in sectors where women are visible, and not in those sectors, such as urban, energy or industry, where the role of women is less obvious.

Some donors give less priority to sectoral interventions, and emphasize more the importance of procedures for *country-level WID guidelines*. Here WID or gender issues are integrated into the country programme/frame documentation included in negotiation procedures between the donor and the recipient countries. These are the major decision-making processes for selecting sector, programme and project priorities. The procedures can include either country-specific WID profiles and plans, or the integration of WID into national plans of action. Finally, there are *WID project guidelines* that provide detailed instructions as checklists to ensure that the project planning cycle integrates gender.

Ultimately, commitment to the integration of WID into the planning process has to be measured. The purpose of *monitoring procedures* is, therefore, to gauge the extent to which each of these six procedures has been implemented. To date, monitoring procedures have focused on the implementation of Plans of Action through such measures as progress reports and 'built-in' monitoring procedures. But progress has been very slow, with the most important problem being the lack of adequate or appropriate indicators. Here it is also important not to conflate the need for two different sets of indicators: first, indicators that assess the implementation of procedures; and secondly, indicators that assess or measure the impact of interventions on the situation of women in developing countries. In the latter case a well-known problem is the lack of disaggregated baseline data against which to measure impact.

One problem in monitoring changes in operational procedures is the difficulty of developing adequate quantitative measurements of process. The DAC evaluation, for instance, comments that although all OECD members have adopted action plans, very few have integrated monitoring procedures to review the process of WID integration in their agencies. Measures, both quantitative and qualitative, are needed to decide benchmarks for WID integration activities. These do not necessarily have to be onerous to develop or cumbersome to use. Measurements such as the numbers of times aid officials raise WID issues in donor exchanges, or the inclusion of WID in the terms of reference, provide simple but innovative indicators.

USAID is one of the few agencies to undertake a rigorous evaluation of the implementation of its WID policy. Undertaken in 1987, its conclusions were both depressing and salutary. This is particularly the case, given the political mandate and impressive budget invested both in increasing institutional capacity and in the development of detailed planning procedures. The evaluation concluded that WID was operating against a range of constraints. These included the attitude of USAID staff, which was one of tolerant indifference. Although lip-service was given to the importance of WID, this was seen as a reaction to special-interest politics rather than a serious concern for development. A lack of regularized systems and procedures revealed that it was the initiative and enthusiasm of individuals that had been the most significant variable influencing the degree to which USAID programmes and projects incorporated WID concerns. Overall, the findings were summarized as follows:

> AID's Women in Development Policy is not being implemented fully or vigorously, and there is little enthusiasm and few incentives for doing so. Without meaningful Agency-wide acceptance of responsibility for implementation of this policy, the efforts of WID and its agents in support of the policy will be marginally useful at best.
>
> (Development Alternatives 1987: 76)

WHAT HAS GONE WRONG? CONSTRAINTS IN OPERATIONAL PROCEDURES

Evaluations such as this raise the question as to what has gone wrong with operational procedures designed to translate policy into practice. USAID's response to this indictment was to develop a new, more complex set of guidelines for incorporating gender into its activities. This was called the Gender Information Framework (GIF), in which both the language and the concern have changed from women in development to gender and development. While it is still too early to see if this set of new tools has changed USAID's practice, the importance of this initiative lies in what it reveals about the agency's perception as to the constraints in the implementation of policy. Above all, these were identified as being of a technical nature emanating from inappropriate or inadequate planning procedure. The solution was to develop 'better' or more appropriate procedures. But is this apparent 'failure' due to technical constraints, or are there also fundamental political constraints operating? The fact that donor agencies have developed so many different procedures makes it important to identify the opportunities and constraints in the implementation of policy.

Is 'symbolic' policy the problem?

It is useful to start by asking whether the problem is that, from its formulation, the policy concerned is essentially 'symbolic'? One reason why policy so often is not carried out relates to the fact that it can easily be formulated without the intention to carry it out at all. It has been argued that effective implementation requires 'unambiguous policy directives' (Sabatier and Mazmanian 1979). 'Perfect' implementation then requires 'complete understanding of and agreement upon, the objectives to be achieved' (Gunn 1978: 173). In introducing the concept of 'policy message', to embrace both the substance of policy and the way it is communicated, Hambleton (1986) argues that there are limits to the degree of clarity that can be expected in the policy message. He identifies a number of sound reasons for policy ambiguity.

First, although it may be easy to specify policy standards and objectives, usually it is extremely difficult to show clear performance indicators or targets. This is particularly true when policy is complex or extensive in its goals. Policies to improve the status of women present a classic example of this case, with globally relevant indicators of status so difficult to identify. Secondly, ambiguities may result from uncertainty, which occurs when understanding is imperfect, or when policy-makers have little control over those carrying it out. In such cases the temptation is to leave policy vague. Again policy objectives to 'bring women into development' provide a widely experienced example of this phenomenon. Thirdly, and of greatest relevance to gender policy, ambiguity may be fostered deliberately by policy-makers. Their intention is to conceal conflicts of objectives between the different actors involved, thus leaving room for manoeuvres, negotiation and renegotiation.

Finally, some policy messages are intended to be heard but not acted on. These policies are not intended to achieve real change but simply to provide 'symbolic reassurance' that something is being done. Symbolic policy is one of the biggest problems faced by those working with a WID agenda. This can be the consequence of policy formulated under outside pressure placed on countries by donor agencies to integrate WID into their development programme. It can also be the consequence of policy formulated without realistic objectives by those unwilling or unable to recognize that policy implementation always involves some kind of compromise. For instance, it could be argued that in many countries the UN Nairobi Forward Looking Strategies are perceived as symbolic policy.

**Are separate WID policies, programmes and projects the problem?
Changing programming approaches from 'targeting' through
'integrating' to 'mainstreaming'**

Procedures to operationalize WID concerns can focus on WID-specific
interventions at policy, programme or project level. Or they can emphasize
'integration' or 'mainstreaming' of gender into existing policies, programmes and projects. This raises the question as to which are the more
appropriate planning procedures. Anderson and Chen (1988) argue that
'programming approaches' to WID can be classified in the same way as
institutional structures, described in Chapter 6. Targeted interventions are
obviously those specifically directed at women. This means that their express
objectives are to meet practical or strategic gender needs, with their budgets
allocated entirely for this purpose. This approach has been justified as
necessary to overcome the gender-blindness that has excluded women from
the benefits of development.

In contrast to this is the 'integration' or mainstream approach in which
women as much as men are taken into account in all stages of the planning
cycle. The rationale for this approach, according to Anderson and Chen
(1988), is that every project activity affects all segments of economy and
society. Consequently, development efforts should take account of those
effects on women. From their review of the relationship between different
programming and institutional arrangements, they conclude that there is no
internal logic that dictates the connection between one structural arrangement
and its programmatic thrust. Both institutional and programmatic approaches
have been successful sometimes and have failed completely at other times.
Each has strengths, but each has also encountered pitfalls.

Despite this conclusion, there is nevertheless an interesting relationship
between different policy approaches to WID and different emphases in
programming (see Table 7.2). Changing positions in this debate have coincided with developments in policy approaches to WID, which appear to have
determined not only the type but also the level of intervention. While the
welfare approach targeted 'vulnerable' groups that happened to include
women with interventions at programme and project level, the equity approach focused on integrating women into development with interventions
primarily at policy level. The anti-poverty approach that quickly replaced it
represented not only a toned-down version of equity but also an important
shift to a different level of intervention. These were the women's projects,
so aptly described by Buvinic (1986) as small-scale, situation specific, and
using limited financial and technical resources. Women, many of them
volunteers with little technical expertise, carried out such projects intended
to benefit only women. The golden era of women's projects was the

Table 7.2 The relationship between WID policy approaches, programming
approaches and levels of intervention

WID policy approach	Programming approaches	Level of intervention
Welfare	targeted	programme/project
Equity	mainstream	policy
Anti-poverty	targeted	project
Efficiency	mainstream	programme/project
Empowerment	mainstream and targeted	policy/programme and project

1970s and early 1980s, associated with the global networking of the WID
concept. However, the anti-poverty approach and women's projects have not
disappeared.

For many agencies, the shift to mainstreaming has been a direct reaction
to the manner in which separate women's projects appear to have further
marginalized women. A common problem identified in government planning
procedures has been for important ministries, such as agriculture, to cease
allocating resources to women-specific projects. This was done on the basis
that the Women's Bureau had their own budget. Equally, as Buvinic (1986)
has argued, resources for poor women are often accessible only if there is not
a cut-back in investment for poor men. This has resulted in a marked
preference for cheaper, non-threatening welfare projects that do not have the
potential to redistribute resources from men to women.

For some agencies, mainstreaming has also marked an apparent epistemological, if not ideological, shift from women and development to gender and
development. Again this has also been accompanied by a semantic shift in
emphasis from 'integrating' to 'mainstreaming'. Thus, in 1982 USAID
WID policy stressed that '*gender* roles constitute a key variable in the socio-
economic conditions of a country' (USAID n.d.: 1).

Above all, the shift from targeting to mainstreaming must be associated
with the increased popularity of the efficiency approach. This identifies
women as the most important under-utilized resource which programmes and
projects must incorporate for more effective and efficient development.
Reflecting the World Bank's position on the efficiency approach to WID is
its Operational Issues Paper on the Forestry Sector, in which Molner and
Schreiber explicitly state:

Women are key actors in the forestry sector throughout the developing
world. Ensuring their direct involvement in forestry projects, both as
beneficiaries and participants, can: a) ensure that projects achieve their

immediate purposes and broad socio-economic goals, and b) maximize returns on investment in this sector.

(1989: i)

The relationship between policy and programming approaches to WID and GAD differs. Separate projects are associated with a welfare or anti-poverty approach, and integrated projects with an efficiency approach. In reality, however, successful implementation relates more to recognition, with adequate budget and staffing, than to whether it adopts a separate or integrated approach. A women's project can be identified as a mainstream project if decision-makers give it priority and an adequate budget (Levy 1989).

Simultaneously, the fact that it is a mainstream policy or project does not a priori mean that there cannot be special components for women within it. For instance, a project targeted at both men and women may require specific women's 'components' to ensure that women participate equally with men. Often agencies do not clearly distinguish between a women's project and a component for women in a mainstream project.

The World Food Program (1987), in their General Guidelines state that a 'women-only' project or a 'women's component' can be normally justified only as a bridging strategy to bring women to a threshold level from which they can enter mainline activities equally with men. One problem here is the failure to recognize women's need to balance their triple role. This means that unless men also take on reproductive activities, no amount of 'bridging' can bring them up to men's threshold. The most obvious women's components in mainstream projects is child care. In reality, this is a family need but is identified as a practical gender need of women because of their reproductive responsibilities (Levy 1989).

Since the mid-1980s most donor agencies have adopted the policy of integrating women into mainstream programmes and projects, as shown in Table 7.1. They use both WID-specific procedures (such as WID-specific sector guidelines) and mainstream procedures (such as integrating WID criteria in sector guidelines). SIDA, for example, has what it terms a 'gendered efficiency' approach. This emphasizes the integration of women into existing development co-operation programmes, rather than establishing separate women's projects. The reason for this approach is twofold. First, women already participate in most sectors, and efforts are needed to facilitate a more effective role for them. Secondly, neglect of women in development co-operation programmes results in a negative impact on women themselves, and on development in general. Efforts to involve women more equitably invariably result therefore in better development co-operation (SIDA 1989: 2).

The empowerment approach to WID seeks to intervene at all three levels of policy, programmes and projects, and combines both mainstreaming and targeted approaches. It recognizes that both types of programming are important. Women-focused projects remain critical, especially at the level of small-scale NGOs, because so many poor women can never be reached in mainstream projects. In addition, mainstreaming, unless meticulously moni- tored, can in effect mean men. There is a very real danger that widespread proliferation of gender, rather than women and development may ironically result in women losing out again. It allows policy-makers and programmers to pay lip-service to women through the term 'gender'. Again, monitoring is the crucial issue. As Christian Aid state in their Gender Guidelines:

> We think there is sometimes a special case for projects run entirely by women for women ... [these] can enable women to increase their level of self-confidence and organizational skill. This is usually more difficult to achieve in mixed [male/female] groups, where there is a tendency for men to dominate – even within groups which are predominantly female. Support for women only projects is justified and should be encouraged, although the dangers of marginalization must be recognised.
>
> (1989: 2)

Is the problem the analytical tools?

The fact that the introduction of gender into existing planning procedures has not fundamentally changed the situation suggests that on its own this cannot confront the real problem. However meticulous the procedures developed to integrate gender into the project planning cycle, it is necessary to assess whether they fall into the trap of being what Thomas (1979) identifies as 'contentless' and contextless'. One important criticism of rational compre- hensive planning is that in assuming 'consensus' it fails to recognize conflict, or to provide the operational procedures to confront it. In 'grafting' gender onto an existing planning methodology, the procedures in the planning cycle have not changed. They do not incorporate new stages that include the negotiation of conflict, or participatory debate. Because it assumes that the problem is a technical one, the introduction of additional interventions within the existing framework of procedures is identified as solving the problem. Therefore, another constraint may be the analytical tools themselves.

One tool that has received particular attention is gender diagnosis and analysis, and the extent to which it recognizes subordination. It is helpful to distinguish between the gender analysis of the 'Harvard approach' and the gender diagnosis approach of the gender planning methodology described in Chapter 5. The Harvard approach was designed to create awareness. Its tools are intended to ensure the meticulous documentation of differences between men and women in divisions of labour, as well as in the ownership and control over resources. It assumes that feeding this information into the project cycle, based on rational comprehensive procedures, is sufficient to change practices. Its careful avoidance of identifying the causes of the inequality between men and women means that it does not seek to provide tools to confront conflict in the implementation process. This is not surprising, since it is based on a consensus, rather than conflict, planning model. Anderson (1990), one of those responsible for developing the Harvard approach, recently identified a gap between gender analysis and 'incorporating this analysis in some effective way into actual programme design and implementation'. However, the Harvard approach does not provide any guidelines as to how this might be achieved.

The Harvard approach is closely linked to the efficiency approach to WID with its economically deterministic underlying assumption that there is a direct relationship between increased access to resources and employment and increased status for Third World women: '[The] Gender analysis approach ... avoided the ideological tone that worried critics ... the approach demonstrated that the issue of women was an issue of economic efficiency, rather than an issue "only" of equality' (Anderson 1990: 31). Gender analysis, therefore, provides tools for policy interventions to help women to participate more efficiently in the labour market, through ameliorating measures such as child-care provision, or transport. It does not provide the tools for policy intervention to help women to 'empower' themselves to organize and challenge existing relations. Policy which equates increased efficiency with equity focuses almost entirely on practical gender needs on the assumption that increased access to employment will make women equal, or more equal, with men.

The cause of this problem is most frequently identified as 'culture', as illustrated in the following statement from the World Bank WID Division:

> For women, as for men, the ability to realize their economic potential depends both on their human capital – i.e. their health and learning – and on their access to information, resources, and markets. However, women face additional and more intractable barriers to access than do men, because of their mothering role (multiple pregnancies and childcare) and *because of cultural traditions*, sometimes reflected in law or policy, that tend to keep women more home-bound than men and more restricted in their work choices and social interactions. These barriers are worse in conditions of poverty, but they persist even in industrialized countries. They restrict women's access to the information and resources required to respond to economic opportunities.
>
> (1990: 2; my emphasis)

Is the current reticence of agencies in recognizing subordination a wide-spread collusion with the idea that better education, health and employment opportunities will make women equal with men? Or is it a sensitivity towards addressing an issue that in reality is so threatening? Gender needs assessment, as a contextually specific planning tool, provides agencies with the capacity to confront this issue more directly. The distinction then between different gender objectives is clear: meeting practical gender needs through develop-ment projects assists women either to perform more effectively and efficiently the activities they are already undertaking, or to take on new activities in their existing roles; meeting strategic gender needs through development projects assists women to achieve greater equality, and there-fore changes existing roles, which, as ODA has argued, may be considered appropriate in those contexts where 'the status and opportunities of women are inferior to those of men' (ODA 1989: 5).

Most governments are committed to policies of equity and equal oppor-tunities, as endorsed in the UN Forward Looking Strategies. However, along with many donor agencies and NGOs, they remain unable or unwilling to articulate the problem of women's position as one of subordination. As Peggy Antrobus has written so explicitly:

> the problem with the array of guidelines, checklists, and methodological tools that have been formulated has less to do with their technical qualities and shortcomings than with the conceptual frameworks, and paradigms, within which they are situated and by which they are constrained.
>
> (1989: 5)

EXPERIENCES WITH PROJECT PLANNING: IS THE PLANNING METHODOLOGY THE PROBLEM?

Given the impressive number of planning procedures developed to operation-alize gender concerns, it is important to identify how far meticulous gender planning tools can ensure the implementation of gender into the planning process. Table 7.1 provided a checklist of the seven most widespread proce-dures, identifying in each case their purpose and country-specific titles. Of these, one important opportunity for gendering planning procedures is project planning. Many donors still consider this the most important planning procedure for developing countries. Consequently, some of the most inno-vative initiatives to integrate gender have been developed in relation to project planning. The remainder of this chapter will focus, therefore, on the factors influencing the integration of gender concerns into the project plan-ning cycle.

Donors have both adapted and adopted the generalized project cycle, described in Chapter 5, to suit their own particular agenda. Governments, NGOs and donors alike now use the logical framework approach, with different modifications. The extent to which the different phases, or stages, of the cycle are clearly identified, however, varies depending not only on the level of professional planning expertise and financial accountability, but also on the ideological position of the procedures, often relying on 'common good sense'. Even where procedures are specified, such as in the Oxfam *Field Directors Handbook*, often they are guidelines rather than checklists for practice. In contrast to this, larger donors such as USAID and the World Bank have highly systematized procedures identified for all stages of the project cycle.

In addition, several new planning tools have been introduced to integrate gender into the project cycle. Although variations exist, it is useful to divide them into three broad categories:

a) *Checklists and guidelines* are intended to be integrated into all stages of the cycle, although their relevance at different stages of the cycle is unspecified. Guidelines, checklists and impact statements are often used interchangeably to both measure and monitor the extent of women's integration in development programmes. Antrobus provides a useful distinction between *guidelines* as 'indicators for translating policy mandates on women and development into action at the programme and project level'; *checklists*, which 'are more specific and aim to provide a more detailed memory aid, giving conceptual clarification and practical suggestions'; and *impact statements*, which are intended to 'get people thinking about the consequences of their interventions' (Antrobus 1989: 13–16).[6]

b) *General project-cycle tools* comprise detailed procedures identified for each stage of the cycle, sometimes in manual form, and are intended for projects despite sectoral concerns.[7]

c) *Sector-specific project guidelines* are detailed procedures laid out for each stage of the project cycle, and for a designated sector.[8]

Chapter 5 described the gender planning process, identifying the interrelationship between gender planning tools, procedures and components in the planning process. This methodology for gendering the project cycle assumes that at each stage of the project cycle an iterative process occurs which includes gender diagnosis, gender consultation and participation, the identification of gender objectives and an entry strategy, followed by monitoring and the inclusion of appropriate institutional structures and training procedures. In theory this is fine; in practice it is more complex. Therefore the following description of the gendering of the project cycles identifies the problems encountered in attempts to incorporate gender, as well as empha-

Table 7.3 Checklist of current interventions to 'gender' the project cycle

	Stage	Important interventions
1	Identification	– policy direction – identification of WID policy approach (WID/GAD policy matrix) – targeted or mainstream intervention
2	Preparation (a) definition of target group (b) identification of gender objectives	– gender diagnosis – gender roles identification – gender needs assessment
3	Design (a) personnel (b) socio-economic feasibility studies	– staff gender training – gendered terms of reference for staff and consultants – mechanisms to ensure women and gender-aware organizations are included in planning process – gender needs assessment – gender disaggregated data over allocation and control of resources
4	Appraisal (a) mission personnel (b) appraisal studies	– gendered terms of reference for consultants – inclusion of gender expert – gender staff training – gendered cost-benefit analysis to include women's 'invisible work' – inclusion of women in staff gender training
5	Ratification	– entry point for dialogue? – staff training on gender-awareness of issues
6	Implementation (a) agency and staff (b) target population	– staff gender planning training – gendered terms of reference for staff – gendered composition of agency – clarification of women's role in participatory projects
7	Monitoring/evaluation	– gendered terms of reference of consultants – staff gender training – team composition

sizing the particularly important interventions at different stages of the cycle. It draws on the comparative experience of a number of institutions that have tackled this planning problem (see Table 7.3). Although stages are discussed separately, they are of course highly interrelated.

Project identification – policy direction

The introduction of the issue of gender into project identification, the first stage in the cycle, determines from the outset the project's orientation.[9] This stage is concerned with relating development goals and resulting sectoral priorities to appropriate projects. These not only have to fit into, and support, a coherent development strategy, and meet sectoral objectives, but they also have to be considered suitable by both the implementing agency and the donor.

The issue of gender is important in two ways in project identification. First, at a general level, gender, along with class, can be identified as an important determinant of development goals in terms of broad sectoral resource priorities. For example, donors often give priority to large-scale infrastructure projects such as dams, hydro-power or electricity, on the basis that their impact will 'trickle down to reach the poor'. Yet the low-income population may not be able to pay for the water or electricity. So such projects, although generally assisting the country's 'development', more specifically meet the needs of factory-owners or upper-income households. Since it is women in the low-income households who take primary responsibility for water and fuel collection, such projects are unlikely to meet their needs. The fact that project identification can predetermine women's participation in projects has been widely recognized. CIDA, for instance, in their 1986 Project Identification Memorandum (PIM) state:

> The project identification stage involves the selection of projects that are compatible with CIDA's aid framework (including the WID Policy Framework) and contribute to sectoral and country strategy. Certain projects have greater potential to influence the lives of women. Provided they are compatible with other CIDA objectives and those of the recipient, such projects should receive preference.
>
> (1986a: 6)

At a specific level, gender is important in terms of the type of project selected. An important distinction exists between women's projects and mainstream or integrated projects, as discussed in the previous section. In reality, the debate between mainstreaming or targeting is critical throughout the project cycle. It affects choices on issues such as staff selection, training procedures and institutional structures. Therefore, a priori, gender must be recognized as a policy priority. As a generalization, projects targeted at those sectors – basic needs provision, small-scale production and farming and traditional food production – have greater potential to reach low-income households.

Project preparation or formulation

The recognition of gender in project identification does not ensure that the project is gendered. It is in the project preparation stage, when target groups are defined, project objectives identified and the project designed, that most has been achieved in integrating gender.

Definition of target groups: gender diagnosis, gender roles identification and gender needs assessment

In principle, it is now acknowledged that gender must be integrated into target group definitions. SIDA (1989), for instance, has argued that the use of collective terms within development co-operation, such as 'families', 'the poor', 'farmers', 'slum dwellers', 'target groups', 'rural population' and 'small-scale entrepreneurs', has negative impact for the integration of women. It is therefore necessary to break down these terms by gender. In practice, however, gender is still not usually included at the design stage – above all, because most planners still lack a gender planning methodology.

Particular problems often occur when planners, overnight, as it were, are informed that 'women must be included'. If, because of 'political' pressure for endorsement purposes at the appraisal stage, the project is required to 'include' women, the document is often simply amended with the phrase 'and women' added to each appropriate stage in the cycle. This happens when an agency funds a project designed by another, or when gender-blind projects are sent for appraisal to WID or social advisers. The fact that there is no logical rationale behind such changes means that usually the project does not in practice become 'gendered'. The most blatant example of this practice occurs when existing project documents add the phrase 'and women' in a different type-face.

Yet an even more common confusion is the illogical identification of gendered and non-gendered categories in the same target group. For instance, 'farmers, landless labourers and women'; 'women and resource poor-farmers'; 'farmers, including farm women and agricultural labourers'. Are all such unspecified project target groups in reality men? When monitoring procedures require agencies to 'count' how many projects include or 'reach' women, lack of accurate gender disaggregated data can result in the under-recording of resource allocations to women. FAO, in a recent analysis of three departments' regular programme activities concluded that '68 percent of projects were not explicitly directed at women to any degree' (FAO 1987: 1). Because this figure only included projects in which the word 'woman' appeared, it presumably missed projects in which 'resource-poor farmers' or 'landless' happened to include women.

Despite constraints such as these, gendered target group identification now has elaborate procedures. Gender diagnosis was described in Chapter 5. However, other methodology has been variously termed 'gender analysis' (USAID), 'gender investigation' (World Bank) or a 'gender perspective' (SIDA). One detailed procedure is that of USAID. It comprises four different techniques: the disaggregation of data by gender; gender distinctions in terminology; the inclusion of explicit strategies to involve women; and the use of gender-disaggregated benchmarks in monitoring and evaluation. To ensure its adoption, USAID has developed the Gender Information Framework (GIF). This itself consists of two components, a *gender analysis guide*, and a *document review guide*.

Based on the Harvard case-study methodology, the gender analysis guide outlines a two-step gender analysis process. First, it identifies where gender might intervene in social and economic production systems (in terms of allocation of labour, sources of income, financial responsibilities, and access and control of resources). Secondly, it analyses the implications of gender differences for project design. These are identified in terms of key differences between men and women's constraints and opportunities in development activities (relating to labour, time, access to credit, education and training, skills and knowledge).

SIDA's (1989) gender perspective identifies two analytical tools. First, *gender-specific target group analysis*. As factors for consideration, this identifies household composition, the division of labour in terms of the different roles of men and women at household and community level, and the responsibility, access and final control over resources at household and community level. Second is *gender-specific impact analysis*. In projects where women may not be considered the 'target group', such as a hydro-power project, there can still be both negative and positive impacts for women, which a gender-specific impact analysis can identify. USAID's GIF and SIDA's target group/impact analysis both provide examples of specific tools intended to ensure the provision of gender-disaggregated data in all stages in the project cycle. The USAID Document Review Guide explains the importance in project/programme design and evaluation, to disaggregate data by gender whenever possible, to incorporate gender considerations and to include decision points in project implementation to allow project adaptation as new data become available.

It is important that gender diagnosis is what it says it is; the diagnosis of both men and women, and the relationship between them. Projects can fail equally if they ignore men in target group identification. SIDA cites the example of health projects, where inputs in child care, nutrition, family planning and care of pregnant women had limited impact if directed solely at women. Men often have ultimate authority with the family. Consequently, even if the women are well informed, they often cannot make changes to family eating habits unless their spouses have sufficient information to agree. For instance, to grow different crops requires that male farmers have the necessary information to make the recommended changes.

Identification of project objectives; gender needs assessment and gender objectives

The integration of gender into project objectives requires detailed specification of the changes which the project needs to achieve for both men and women. The basic issue here relates to the assessment of gender needs, and the extent to which the methodology distinguishes between practical and strategic needs. CIDA, for instance, in their PIM document provide a guide of eleven issues to ensure that their objectives address key aspects of WID. They state that projects should be examined in terms of whether they

- lead to jobs, skills, training, improved markets, more public services, housing, food production or other gains that directly benefit lowest income groups, in which women frequently are found in the largest percentages;
- facilitate access to and control over resources by poor women, including women heads of household (1986b: 100).

One constraint of a checklist such as this is that it does not prioritize needs into a 'cause and effect' hierarchy, or state the likelihood of their being achieved. Thus it identifies practical gender needs relating to jobs and skill training alongside strategic gender needs related to controlling resources. As ODA (n.d.) points out, project objectives need to identify whether they recognize the distinction between different types of needs. Christian Aid recognizes that objectives to reach practical gender needs automatically do not reach strategic gender needs. These often require specific, often highly sensitive, interventions in their own right. In its Gender Guidelines, it shows its priorities in terms of gender needs:

> Our task is not simply to support the involvement of more women or to fund more women's projects, but to equip women with the means to challenge and overcome their constraints. We should address ourselves to both the practical and strategic needs of women. We should support projects that enable women and men to understand the structural inequal-

ities which adversely affect the participation and development of women, and that include action to address those inequalities.

(Christian Aid 1989: 1)

Project design and gender consultation and participation

Project design includes questions relating not only to what is designed, but also who designs, and how it is designed, all of which involve issues relating to gender. The procedures discussed under project design, which relate to personnel and feasibility studies, also relate to later stages in the project cycle. These include *project personnel, gender training and gendered terms of reference*.

One difficulty in planning relates to the division of labour at different stages in the project cycle. Staff identifying project objectives may not design the project. Unless both are gender-aware, project objectives may not be taken into account in the design phase. Agencies commonly separate out the design of 'technical', 'economic' and 'social' components of projects to different experts. This encourages 'technical' experts to leave the social development issues to the socio-economists, or anthropologists. In order to solve the problem of gender-blind personnel, most donor agencies have integrated training. The extent to which this helps technical staff depends not only on the relevance and content of training, but also on the policy approach to training itself – issues to be examined in detail in Chapter 8.

The fact that training provides no guarantees that staff will be sensitive to issues in practice has resulted in the operationalization of a further critical planning tool. This is the development of clearly gendered terms of reference of all project staff. This identifies the gendered information and analysis required to ensure gendered interventions at each relevant stage of the project cycle.

The socio-economic feasibility study

The timing of feasibility studies is often critical to project success, with severe difficulties experienced in making changes later in the project cycle.[10] The socio-economic feasibility study can provide the first critical entry point for gender issues in the project cycle. To ensure they include gender, many agencies have redefined their project formulation stage to include social studies.

At the project design stage it is the gender-specific potential inputs and outputs that the feasibility study needs to identify. Molner and Schreiber (1989) identify four specific issues: pre-project benefits likely to be forgone by women and their households (with special attention to households headed by women); work-load implications for women; probable gains for women; and differences and potential conflicts between probable gains and losses for women and those anticipated for men, households in general or for the community as a whole. CIDA (1986a) identifies the project feasibility study as a synthesis of separate studies that deal with four essential aspects of the project, which they identify as technical, managerial, socio-economic and financial. CIDA states that the feasibility study should identify the institutional and other changes needed to facilitate, increase, accelerate and retain the access and control over resources of the sector by women. Information on the degree to which women have control over resources can be used to assess the extent to which they are likely to retain control over benefits of the project. It also assists in identifying potential conflicts that may occur if project design challenges existing roles, benefiting women and disadvantaging men.

Like USAID, with its socio-economic studies, CIDA uses the Harvard case-study approach. However, it also stipulates that information should be obtained from local people (men and women) and local academics, leaders, women's organizations and NGOs. Equal consultation with and participation of local women and men is essential, although often difficult to achieve in practice. Gender diagnosis also has important implications for project design. For instance, information on women's reproductive role can be used to assess whether the project will have implications for their need to balance their triple role. Buvinic identifies a critical design fault in income-generating projects as the underestimation of the difficulty of stereotypical female tasks, and the overestimation of their transferability.

> The typical project involves group activities through which women attempt to apply the skills they have learned for income generation... common wisdom judges that stereotypical Western female tasks are both simple and familiar to poor women in the Third World and are, therefore easily transferable. . . . In reality however, female appropriate tasks are not simple, nor are they as familiar to low-income women as they are assumed to be.
>
> (Buvinic 1986: 656)

Since many issues relating to the feasibility studies are equally important at the appraisal stage, they are discussed below.

Appraisal

Project appraisal is intended to provide a comprehensive review of all aspects of the project and lays the foundation for implementing the project and evaluating it when completed (Baum 1982). Traditionally, appraisal covered technical, institutional, economic and financial aspects of the project, with social and environmental appraisal being a more recent 'add-on'. From a gender perspective, this is a critical stage in the project cycle. For donors it is often the first stage when they formally participate in the project. Therefore, it may be the only opportunity to make any fundamental changes in the framework already developed.

Appraisal mission personnel: gender experts or gendered terms of reference

Donors have increasingly realized the need not only to ensure that all terms of reference for appraisal missions are gender-aware, but also to get gender specialists on the appraisal team. Does the team need a gender specialist if all members are gender-aware? This debate is similar to the mainstream–target debate outlined above. In addition, one specific problem that occurs is the assumption that women professionals will automatically monitor the gender aspects of a project. The greatest resistance to mainstreaming often comes from women professionals. They argue that it undervalues their professional expertise. By default, they become a 'women's' expert, as against a technical expert.

Both NORAD and SIDA, for example, identify as an important priority the appointment of a WID specialist on the appraisal team for both projects and country-sector mission reviews. The task of the specialist is to act as a catalyst. The resident in-country WID officer cannot be expected to take on this responsibility. Their status may not be high enough, and as part of the residential mission they are likely to experience problems integrating into a visiting team.[11] However, specialists are considered to be a short-term strategy; in the longer term SIDA expect to mainstream the issue. A recent example of terms of reference for a SIDA appraisal mission – for continued support to the road sector in a developing country – required it to give attention to increasing the participation of both women and men in the sector support. To facilitate this it required a gender assessment of roles, responsibilities, access to resources, and the resulting needs and problems of both women and men.

Integration of gender into technical, financial and economic appraisal reports

Where gender is recognized at the appraisal stage, it usually consists of a statement about women, added to the social appraisal. This is the standard response to a checklist question about the way in which the project 'affects' women. However, it results in a twofold problem. Not only is the social appraisal rarely authentically gendered. In addition, none of the other 'technically' focused appraisals contains any social dimensions. Yet there are critical gender issues in all the different appraisals undertaken. Although many are sector-specific, analysis of the anticipated project impact in terms of women's triple role provides the basis for identifying effects on them at the local level.

Economic appraisal reports are considered the most important, with cost-benefit analysis utilized to identify alternative project designs, from which the one that contributes most to the development objectives is selected (Baum 1982). One important issue in cost-benefit analysis is the extent to which the invisible work women do in their reproductive and community managing work has an economic value. This has been a particular problem in recent structural adjustment policies where projects to shift production from non-tradable to tradable goods have failed to value women's time as peasant workers (Elson 1991). It is not simply that such interventions can harm women; gendered cost-benefit analysis shows that economic returns are not reduced in projects that specifically ensure benefits to women. In addition, the direct involvement of women can improve the probability of projects meeting their essential objectives. Despite such conclusions, the assumption still common among economists is that 'involving women as direct beneficiaries or participants and devoting special resources to achieve this is peripheral, if not detrimental, to the attainment of central project objectives, and/or requires additional inputs at the expense of overall project focus and project returns' (Molner and Schreiber 1989: 27).

Once those responsible for the technical appraisal recognize gender as a planning issue, specific components relating to project layout and design, location and scale and type of equipment can be identified. Among the most widely cited are: that project location can differentially affect the participation of men and women; that unless the project specifically itemizes women as beneficiaries, land for 'landless' and legalized plots for 'squatters' in both cases reaches men; and that unless the project consults women, prioritization in delivery of basic services is usually those that benefit men (Moser 1987a).

A particular training problem in the technical appraisal is the selection procedure for personnel. Generally this is based only on technical expertise and not on gender-awareness. When the project includes training components, women do not always get equal access to such training, nor do the location, hours and length of training help them to participate. Again it raises the problematic issue of mainstreaming or targeting. Molner and Schreiber argue that

Almost everywhere it will be necessary to educate male staff about women's roles in forestry. In some instances, this may actually be more realistic and cost-effective in the short term than attempting a rapid increase in the deployment of female extension staff. Properly oriented and trained, male staff can effectively consider women's needs and contributions and reach women directly in field work.

(1989: 14)

Not all would agree, however, that the problem is simply one of lack of training, with gender-aware men then being able to solve generations-old problems.

Project ratification of negotiations

In his description of the project cycle, Baum (1982) identifies negotiation as the stage at which the donor and the borrower endeavour to agree on the measures necessary to ensure the success of the project. Such agreements are then converted into legal obligations, set out in the loan document.

This stage is often identified as one of formal ratification. However, the extent to which 'political space' exists for real negotiation around differences of opinion between either donor and recipient, or from different groups within the recipient country, will determine its importance in the project cycle. For some agencies it may be the stage for document ratification, such as CIDA's Project Approval Memorandum (PAM). For others, such as SIDA, there is a negotiation period between appraisal and agreement, which can be lengthy. This stage is then an important gender entry point for shifting the agenda. Since senior staff generally control this stage, the extent to which those negotiating represent women's interest depends on the gender-awareness of participants in the negotiation process.

Implementation

Implementation is the period of construction and subsequent operation of the project. Baum has argued that although supervision is the least glamorous part of the project work, it is the most important. For no matter how well a project has been identified, prepared and appraised, its development benefits can only be realized if it is properly executed (Baum 1982). Given the widespread failure of gender policy in the implementation phase, it is particularly important for gender planning. Several gender-related issues here require specific identification and close monitoring. Many are similar to those raised in earlier stages in the project cycle. At this stage it is important to ensure that gender objectives are carried out in practice. Those most affected, therefore, are the implementing agency and the target group itself.

Implementation organizations and their personnel

So often when the project is one of infrastructure provision the implementing agency does not include social scientists among its personnel. Staff composition is critical not only in terms of sensitivity in understanding the project, but also because of its implications when the target group participates in the project. Too often professionals, both consultants and nationals, bring their own stereotypical assumptions about the role of local women in society.

CIDA (1986b) argues that an important issue concerns how managers or administrators are motivated, trained and/or persuaded to consider women's contribution. This raises the issue whether it is better to use existing organizational structures or create new ones for the implementation stage. Widespread concern at the way highly over-bureaucratized male-dominated institutions fail to reach women has resulted in an increasing focus on local women's and NGO groups at the grass-roots level. With women's groups, the problem, however, is not so much the groups themselves, but their resources and staffing. The widespread use of women as volunteers not only means that 'staff' often have inadequate training. As Buvinic (1986) has argued, it leads to the perpetuation of women's lower status among those implementing projects. Buvinic claims that the use of women volunteers is based on the 'cultural prescription' that work with low-income women should be done by women. She claims that the validity of this belief has gone largely untested in the Third World. Lack of expertise is often a hindrance to project success, while the cultural need to have women staff members interact with women beneficiaries will vary according to the poverty of the group and the task required. It is important, however, not to conflate the problems of untrained volunteers who are also women, with trained, paid women project staff.

Finally, it is important to recruit women professionals into management positions. However, this will help local women only if the former are gender aware. Where the project involves extension or community workers in implementation, the terms of reference must include women in this activity. However, 'women's components', such as crèche facilities, or extra training may be necessary to ensure that women gain access to employment on equal terms with men.

The role of the target group in implementation

When implementation or maintenance involves the target population, women's participation varies with project type. Projects with labour components increasingly use women to provide free or cheap labour. These projects can range from the provision of infrastructure and housing to those allocating handouts such as food, or providing services such as health. Recognition of the triple role is obviously essential, if the participatory component is not simply to extend the working load of low-income women, who participate to get services. As discussed, community management does not mean that women participate in community-level political processes. The failure of implementation agencies to ensure that women participate in decision-making processes has resulted in the creation of separate women's community-level organizations. These ensure that women can express their own choices and make their own decisions.

In its Plan of Operations, CIDA questions whether projects have the institutional capability to support and protect women's interests and, if not, how this can be developed. It raises critical issues concerning an agency's capacity to ensure that it effectively sustains those support structures, created to help women in potentially conflictive situations. Where the project is simply reinforcing gender roles this may not be problematic. However, where objectives have included strategic gender needs of women, or where women themselves have identified these during participation, the project often cannot sustain itself unless an adequate support structure exists. Finally, it is important to recognize that women in the target population are not a homogeneous group. For instance, as Buvinic (1986) has commented, women who are better off and do not need to work for a living self-select themselves for project participation. In contrast, those who head households, and who often are the poorest with the severest time constraints, exclude themselves from projects that require time for group discussion, participation and voluntary labour.

Monitoring and evaluation

Although monitoring is sometimes identified as a separate stage in the project cycle, increasingly it is identified as an ongoing activity accompanying the implementation process. Ideally, a project should establish an ongoing data-collection system for monitoring purposes. However, this is only realistic for large-scale or prototype projects, but is too ambitious for most projects. Here mid-project missions or on-ground monitors can provide essential feedback during project implementation at a reasonable cost (see CIDA 1986a). However, unless the project already has undertaken gender diagnosis and integrated gender, it is unlikely to include it at this stage. Yet monitoring of women's participation can be ensured if all the data found during the project implementation is collected and analyzed on a gender-disaggregated basis. In addition, special studies and surveys may be required, for which adequate financial provision should be made. Molner and Schreiber (1989), for instance, identify the collection of data on women's own perceptions of the project's impact; on conflicts between men and women's interests in project outputs; on the extent and distribution of project 'costs' and project-induced 'losses' among various users and beneficiaries; and on the needs, demands and constraints of different users over time.

Evaluation is the final stage in the project cycle, providing an insight into factors that contribute to project success or failure. As with monitoring, it is essential to ensure that the indicators developed are adequate and relevant to the evaluation of gender. Another constraint may be the lack of adequate resources. CIDA, for example, only evaluates a few projects, because of resource constraints. It has recommended, therefore, that priority should be given to projects that address WID concerns.

The first important step is establishing the reasons for evaluation. Projects are evaluated against their original objectives, so, if these were not gender-aware, neither is the evaluation likely to be (Levy 1989). However, the evaluation of gender in the project can ensure that the second phase includes gender. SIDA tries to emphasize socio-economic aspects within evaluations.

To ensure objectivity and independence, evaluations are not generally undertaken by the operational agency but by outside consultants or separate departments set up for this purpose. This makes issues such as the terms of reference, and consultant selection procedures, discussed earlier, of particular importance. Again it is important that the evaluation team includes a gender specialist, an accepted professional, not the in-country WID officer. Another argument, made by CIDA, is the need to involve host nationals in the research process. Not only does this help the evaluation process, but it also ensures a transfer of research techniques and expertise to the recipient. Both with consultants and nationals it is preferable to have female inter-viewers interview females, while in certain countries it is imperative. Finally, to ensure that the findings and lessons learned about WID are highlighted, CIDA suggest that the evaluation report contains a separate chapter or section on WID.

CHALLENGING AND CONFRONTING BLOCKAGES: THE POLITICAL AGENDA OF GENDER PLANNING

The previous description of the range of analytical and planning tools developed by various agencies for use in the project cycle illustrates two important issues. First, there are very real differences between the tools and techniques developed to date. At one end of the spectrum are brief, gener-alized checklists, at the other end are complex, detailed procedures integrated into each stage of the cycle. Second, despite the gender approach adopted, there is a broad consensus concerning specific tools, such as gendered terms of reference, staff training and gendered selection procedures. Both these issues have important implications. They illustrate exactly how complex, time-consuming and expensive it can be to gender the project cycle. Each stage requires detailed scrutiny to identify the techniques required, and continual monitoring of their effectiveness once introduced. This, if nothing else, serves to illustrate precisely how difficult it is to operationalize gender planning.

Has the extensive introduction of new procedures changed the situation? Lack of comparative data means it is not possible to assess the progress of different agencies about their planning tools. Nevertheless, it cannot be assumed, a priori, that the greater the sophistication of existing tools the more effectively gender is integrated into the planning cycle. Agencies are naturally reticent about evaluating their own experience, preferring to view whatever changes and shifts that have occurred in the most positive manner possible. However, there is a consensus that there is still a long, uphill struggle before changes in procedures have any substantial impact, both in terms of changing attitudes within their agencies, as well as in fundamentally improving the situation of women in developing countries.

A recent evaluation of UNIFEM's work mainstreaming gender documented the continuing indifference, ambivalence and active resistance that efforts to mainstream women into development policy and programming still encounter. It concluded that 'resistance, born out of prejudice or protective self-interest is a fundamental fact' (Anderson 1990: 44). Even those professionals sympathetic to WID can seriously underestimate the effort involved. The previous chapter cited the case of an NGO head who viewed the establishment of a WID Unit as a 'temporary phenomenon' for two or three years, based on the assumption that this was the necessary time to institutionalize WID. However, after more than five years this has still not been accomplished. The question can then be asked whether the 'lack of regularized systems and planning procedures' is the essential problem in operationalizing WID. Recognition of these problems in operationalizing gender has resulted in the development of several strategies to challenge the blockages in planning procedures.

Operationalizing gender planning: a case study from SIDA

This chapter has identified a number of important shifts. These have occurred not only in the focus but also in the level of intervention. This includes changes from women in development to gender and development, from project to programme and policy level, and above all from 'targeting' and 'integrating' to 'mainstreaming'. Mainstreaming as much as targeting or integrating will fail unless the planning methodology provides procedures to confront blockages. This requires the complex but necessary identification of the fact that planning methodology is both technical and political in nature. This chapter concludes by describing briefly the planning procedures recently developed by SIDA. This donor agency is changing both its institutional and operational planning procedures, through the integration of the gender planning methodology outlined in Chapter 5.[12]

From their experience to date, some important conclusions can be made. First, like many donors, SIDA has recognized the importance of a combined strategy. Its gender policy approach is 'modified' efficiency combined with empowerment. The former is integrated into sector programmes through sector action plans, while the latter relates to specific actions for women through the direct support programme and other special funds for democracy, human rights, the environment and other specific areas such as population and culture.

Secondly, SIDA has recognized that their biggest constraint has been the *lack of an adequate planning methodology*. SIDA, therefore, has simplified and standardized the gender planning methodology to ensure its incorporation into all interventions. Highly complicated procedures, while impressive, are time-consuming and more open to misunderstanding, and on both counts are less likely to be carried out. SIDA, therefore, identify three simple procedures for all planned interventions: the integration of gender into mainstream programmes as opposed to separate projects/components; the identification of responsibilities and roles within SIDA; and the incorporation of gender planning methodology into all normal planning routines.

Thirdly, SIDA recognizes the impossibility of acting effectively on all fronts at the same time and, therefore, prioritizes within its gender strategy. Fourthly, it identifies the preparation stage as the most important stage in the planning cycle. It recognizes that the critical criteria for the selection of entry points through which gender is introduced are 'political', and related to the points in SIDA's planning cycle where there is greatest political space for negotiation. As a result, three stages of SIDA's programme cycle have been selected as entry points. These are identified as the Preparation of New Support, when the donor and recipient undertake detailed negotiation of future support; the Yearly Sector Review, undertaken in each recipient country; and the Evaluation, when it incorporates gender in the next planning cycle. These are only three of many stages in the programming cycle. However, they are identified as providing the greatest leverage for confronting contextually specific issues relating to women's strategic as much as practical needs.

To ensure that the political space is utilized to maximum advantage, three specific 'technical' planning tools – terms of reference, team composition and reporting back – are incorporated in all entry points. Finally, several tools to ensure the mainstreaming of gender are also identified. These include personnel with catalytic roles at both headquarters and DCOs, and adequate funds and resource bases (discussed in detail in the previous chapter), as well as operational tools similar to those identified earlier in this chapter.

Along with most agencies involved in development work, SIDA identifies that an important constraint remains the lack of gender-awareness of colleagues, or the lack of capacity to translate awareness into practice. It has, therefore, placed special emphasis on the development of training. This includes gender specialists, as well as aid generalists within the agency. Chapter 8 examines the extent to which training provides the solution for both operationalizing and institutionalizing gender into planning.

8 Training strategies for gender planning
From sensitizing to skills and techniques

In the past decade training has become identified as a critical component in ensuring that organizations successfully integrate WID or gender planning into their work. A proliferation of new 'approaches', courses and workshops have abounded as different institutions and individuals have sought to fill this need. In many contexts training is now identified as the panacea, the solution to all problems. These can include gender-blindness, rigidly entrenched attitudes and hostility to women's concerns, or plain ignorance of how to integrate the issue of gender into planning. In reality, training is undertaken for different objectives, with both the training methodology and the components of the training package varying accordingly. The purpose of this chapter is to identify the potential and limitations of training to assist in operationalizing and institutionalizing gender concerns into planning practice. How far can training solve problems, and what does it actually solve?

The chapter commences with a brief description of the different types of training courses developed over the past decade. It identifies three principal *training methodologies*, termed *gender analysis*, *gender planning* and *gender dynamics*. It then highlights the different components of a gender-training strategy, drawing on a decade of personal experience. Five fundamental questions – *why* train, *when* to train, *who provides* training, *who undergoes* training and *how* to train – provide the framework for addressing training issues in this chapter. Gender training is a complex, sensitive and sophisticated field of work, conceptually, methodologically and, above all, in its practice. The 'first generation' of gender trainers has already learned through bitter experience that, despite the immense demand for such training, neither 'instant fixes' nor 'quick sales' of guidelines, manuals and packages provide the panacea to gender planning problems. This chapter does not attempt to evaluate different training packages. Its purpose is simply to highlight several important substantive and methodological issues of use for those involved in the development of gender-training strategies.

BACKGROUND: THE IDENTIFICATION OF THREE TRAINING APPROACHES

In the past decade, along with the development of different conceptual approaches to the issue of women and development, and the varying institutional structures and operational procedures to ensure their implementation, different training strategies and techniques have developed. WID/gender training developed primarily as a response to gender-blind practices. However, the different training courses have also provided the 'testing ground' for the intellectual development of different analytical approaches. Consequently, the different approaches to training coincide with the different approaches to WID issues, described in Chapter 4.

Many training packages are now in existence, with a range of training experience undertaken with aid donors, Third World government practitioners and NGOs. A recent conference (May 1991) in Bergen, Norway, provided an important opportunity to evaluate a decade of gender-training experience.[1] Some ninety conference participants reached consensus that of the diversity of experience represented, it was useful to identify three general approaches. Although these have emerged as 'dominant', it does not mean to suggest that they comprehensively cover all methodologies.

Gender analysis training

The first, and by far the best-known and most extensively used, is gender analysis training. This originated in 1980 when the World Bank WID Adviser commissioned a team headed by James Austin, a well-known case-method trainer at Harvard, to conduct a series of workshops for World Bank staff. He brought in three women with WID experience – Catherine Overholt, Mary Anderson and Kathleen Cloud – who formed what is known as the 'Harvard Team'.

The basis of the Harvard approach is gender analysis. This is identified as a diagnostic tool. It consists of a sequential, threefold analytical framework to address the division of labour between men and women and their different access to and control of resources. Case studies are used as the 'pedagogical vehicle through which intellectual involvement is generated' (Overholt *et al.* 1984: xiii). These describe a set of events and provide the available relevant raw data but leave analysis and evaluation to participants. As Overholt has identified, in the real world the case situations do not have one right answer (Overholt *et al.* 1984).

The analytical framework developed for the analysis uses four interrelated components: activity profile; access and control profile; analysis of factors influencing activities, access and control; and project-cycle analysis. In the workshop format, participants use case studies in relation to the analytical framework. With this tool they identify gender-based divisions of labour and access to and control over resources. This enables them to highlight the often invisible nature of women's contribution as well as the differential impact of proposed project interventions on men and women. Repeating the process with different sets of data but guided by the same framework, the intention is to train participants to be able to use gender analysis as a tool in their own work. This shows them that the framework guiding gender analysis is applicable to all situations (Poats and Russo 1989). Over the past ten years the Harvard team has worked closely with, and had an impressive impact on, most North American donors. Institutions such as USAID, CIDA, UNDP and IDRC have adopted, and adapted as appropriate, gender analysis training. All have developed detailed tools based on the gender analysis approach for the training of aid donor personnel both at headquarters and in the field.[2]

Gender planning training

A second approach is *gender planning training*, which is linked to gender planning. In collaboration with colleagues at Gender and Planning Associates, I was responsible for the development of this approach. This began in 1984 with a course, 'Planning with Women for Development', initiated at the Development Planning Unit (DPU), University College, London, in collaboration with Caren Levy. This was a three-month short course specifically targeted at Third World planning practitioners (Moser 1986). In 1987, on moving to the London School of Economics (LSE), I developed a one-year academic course, 'Gender, Development and Social Planning', based on the same conceptual rationale of gender planning.

Simultaneously, UK-based development NGOs expressed a need for short one-to-three-day courses. So with a VSO officer, Sukey Field, I developed a one-day course. Since then NGOs such as Christian Aid, Oxfam and VSO have used this short course extensively (Christian Aid 1987). In the past five years the same course format has been developed further for training courses varying between one and five days. Donors such as ODA, SIDA, NORAD and the Commonwealth Secretariat have used these, with trainers in each case developing contextually specific training strategies (Hannan-Andersson 1991). One basic premise of this methodology is that it contains components for the training of trainers. Consequently, several Third World participants from both the DPU and LSE courses have developed their own training courses based on this methodology for use with governments and NGOs. One interesting example is that of the Mujer y Planificacion del Desarollo Programme (Women and Development Planning) in Lima, Peru, that runs four different categories of training activities (Fort 1991). In the past two years Caren Levy, in particular, has further developed gender planning procedures such as gender diagnosis and entry strategies. Finally, this methodology has been used to advise such institutions as FAO, ILO, UNIFEM, CUSO and the Africa Technical Department of the World Bank on the development of their training strategies.

Because the Appendix outlines the gender planning training approach in detail, this chapter only highlights the most salient points. The basis of this approach is planning, rather than analysis, and gender planning rather than planning for women in development. The purpose of the training is, therefore, to provide tools, not only for diagnosis, but also for translation into practice. The methodological tools identified simplify the complex theoretical feminist concerns discussed in Chapter 2. These relate to the productive, reproductive and community managing roles of women, to decision-making within the household and to the nature of women's subordination. The purpose of simplification is to translate these concerns into specific interventions in planning practice. Tools such as the triple role, gender needs assessment, the WID/GAD matrix and gendered participatory planning procedures help planners to undertake gender diagnosis, define gender objectives and identify gender entry points. In addition, it assists them to recognize the constraints and opportunities in institutionalizing and operationalizing gender planning within their own organizations. These tools provide a common language that allows practitioners from different disciplines, of different persuasions and positions on WID/GAD issues, to communicate with each other non-threateningly. The simplification of complex realities has obvious dangers. However, it also has advantages in that these realities can be quickly grasped and translated for utilization in planning practice.

Training in gender dynamics

Gender dynamics is a third, and very different, training approach. It differs fundamentally from the first two in that it comes mainly from the training experience of Third World grass-roots organizations, as against First World GAD researchers. In addition, its constituency is Third World practitioners, particularly women in NGOs, rather than government, or First World development personnel.

To date this approach has not been well documented. It comprises several indigenous, highly participatory, innovative and flexible methodologies designed to 'empower' women to recognize, analyse and address gender issues at the grass-roots level. Interpersonal skills, that engage 'not only the mind but the heart' (Aklilu 1991: 25), provide the basic training technique.

In a recent description of this approach in the Philippines, Theresa Balayon describes it as follows:

> gender dynamics is a consciousness-raising seminar that discusses questions relating to the nature, origins, extent, effects of, and alternatives to, gender bias... [to] identify gender bias at home, at work, and in society-at-large; interpret some experiences in the light of feminist values and principles; reconcile various points of view in the women's movement; correct sexist patterns of thinking, speaking and behaving; and integrate gender-fair values into one's personal and professional plan of action.
>
> (1991: 1)

As the term suggests, 'gender dynamics' is based primarily on 'interactive' lectures, role play and interpersonal dynamics. In some contexts, additional tools have included the use of popular theatre, such as the work of Sistren in Jamaica.

THE COMPONENTS OF A TRAINING STRATEGY

Before undertaking gender training, a training strategy is required. This contains a number of components, which can be addressed in terms of a series of interrelated questions. The following discussion raises issues relevant to all three training approaches identified above. However, in the main it describes the most important components of a gender planning training strategy (drawing on experience in this field, particularly with donors and Northern NGOs).

Why train? The objectives of different training approaches

Training does not occur in a vacuum. As Anderson (1991) has identified, training approaches and their effectiveness are influenced by the social, economic and institutional context in which they occur. Nevertheless, the basic assumption on which all training rests is that this is a powerful 'transforming' tool through which people learn new attitudes, knowledge or skills. Once acquired, this will make them become more effective at what they do (Anderson 1991). The results of a recent survey of gender training show a clear agreement that the long-term goal of training is to achieve a gender-equitable society, and is captured well in the goal statement made by the East and Central Africa Training Institute (ESAMI):[3] 'To initiate change at the individual, organizational, social and policy levels which encourages gender responsive sustainable development' (Cloud 1991: 3).

When training is identified as important simply because it is number six on the action plan, or because the Nairobi Forward Looking Strategies endorsed it, this often results in unfulfilled expectations. It then reinforces lack of legitimacy of WID personnel, and negative attitudes of staff towards WID issues. It is the confusion or lack of clarity in many organizations about the purpose of training, therefore, which needs to be addressed.

The purpose of training, however, may differ widely from a broad concern to create a general awareness of gender issues to the specific need to impart skills and techniques. If training is undertaken because most staff are hostile to WID (in which case its purpose must be sensitization) then its objective is to make gender-blind staff aware of gender. As Staudt (1990) argues, it is important not to forget that many agencies continue to identify the conventional gender ideologies of staff members, and a lack of acceptance of WID as a legitimate professional concern, as the strongest reasons for 'bureaucratic resistance'. Alternatively, if staff are interested and sympathetic but lack the necessary tools to integrate gender into their work, the purpose may be to provide them with the appropriate methodology. This does not necessarily mean that staff will develop a 'feminist' orientation to the issue; their concern could be one of efficiency, not of equity. Of importance here is the WID/GAD policy matrix that can be used to inform staff of the rationale underlying different policy approaches. It is important, therefore, to clarify the objectives of training. These have direct implications both for the development of a training strategy, and for the components of the training methodology. While many training courses combine a number of different objectives, it is useful, nevertheless, to start by distinguishing between four basic objectives.[4]

First, there is training in *sensitization* or *awareness-raising* to the importance of women and development and gender and development issues. The objective of this type of training is to introduce participants, who can be identified as 'gender-blind', to both women and gender as variables in the development process. Most early gender-training courses were preoccupied with this objective. Although the situation has changed considerably, Cloud's (1991) review concludes that this remains the main domain of gender training courses, as the first of several stages in the training process.

Secondly, there is training in *skill transfer in gender analysis and diagnosis*. The objective is to impart the necessary skills to gender-aware participants to enable them to undertake gender analysis and diagnosis. While the specific tools may vary, their purpose is to enable participants to analyse a number of variables. These include the gender division of labour, access to and control over resources, gender needs assessment and the underlying policy approaches to WID. These tools can be used in gender diagnosis of concrete development contexts. They also can be used to appraise or evaluate the ongoing policies, programmes and projects of many different agencies working in various sectors. Gender analysis training, described above, focuses particularly on this objective, with the development of a highly sophisticated methodology.

A third objective of training is the *translation of skills into planning practice*. This ensures that participants develop the capacity to translate their 'theoretical' apparatus into 'practice', through its implementation in ongoing work. It aims to achieve the integration of gender planning methodology into both the institutional structures and the operational procedures of the organization in which participants work. By its very definition, gender planning training places greatest emphasis on this objective.

Finally, there is training in *motivational factors*. Its objective is to motivate participants to 'do the job'. The extent to which this is necessary varies. It depends on whether constraints in gender planning relate to personal attitudes or lack of professional skills. In addition, approaches to training vary, depending on the extent to which confrontation is considered appropriate. Gender dynamics training identifies motivation as the primary constraint and focuses on this objective.

Underlying the different training objectives are profound questions concerning the extent to which they address the issue of subordination. Associated with this question is that of the underlying policy approach to WID. Professionals differ widely in their perception, or recognition, that gender training is confrontational. The basic premise of feminists that the 'personal is political' presumes such issues to be highly contentious for those who do not agree that women are subordinated to men. Their assumption is that training on such an issue must be highly susceptible to confrontation. Obviously, training on planning for WID very neatly avoids this problem. Because it focuses on an efficiency approach, it provides the rationale to deal with women on their own, as a separate category. However, even for those who recognize the issue of subordination there are useful tools that can help in reducing confrontation. Since highly aggressive challenges can so easily destroy the effectiveness of a training course, these are obviously important. In gender planning training courses, for instance, one successful way to diffuse tension is to distinguish at the outset between the '*professional*', the '*personal*' and the '*political*', and to place particular emphasis on the fact that gender planning training concerns the '*professional*'.

The logic is the following: if the 'personal' refers to the way in which individuals understand and analyse their personal relationships between men and women in their own society, the purpose of training is not to change such personal beliefs. Equally, if the 'political' refers to the feminist agenda in the individual's society, the training is not intended to challenge or change political stances on this issue. An emphasis on the 'professional' therefore assumes, for strategic reasons, an a priori awareness of the importance of women's role in development. It emphasizes focusing on the 'technical skills' required to improve professional competence. This allows participants to accept without hostility such tools as gender roles identification, gender needs assessment and the WID/GAD policy matrix. Ultimately, the distinction between practical and strategic gender needs is a critical tool that helps participants recognize the interrelationship between the political, professional and personal. Indeed, the diffusing of tension is an inherent part of the tools themselves. In demystifying the widespread belief that all interventions to assist women are 'feminist', it allows participants to 'put on the back burner' those interventions considered inappropriate. Yet it also allows the beginnings of a re-examination of 'personal' as well as 'political' positions on the issue of women's subordination.[5]

It is important to note that this mechanism to 'dilute' confrontation is appropriate for training those in the 'international community' working on Third World issues. One unrecognized complication in gender training is the important distinction between practitioners working on development issues in their own environment, and those outside working in a donor capacity. For the latter, this training does not directly address gender concerns of their own society. Because it is, in a sense, one step removed from the reality of their personal lives, such training is easily identified as professional training. For professionals working in their own society this is not so. This, I would argue, is the reason why training in gender dynamics is more widespread within Third World countries. Third World trainers consider it more appropriate to confront such issues directly than do those developing training methodologies for use in the international community. As Cloud has commented,

> Women representatives in countries with an expressed commitment to justice are in a position to press for greater equity in every area of national behaviour ... it is very difficult for outsiders, particularly male outsiders, to suggest that the gender arrangements of another culture are unjust, especially those within the family.
>
> (Cloud 1991: 4)

Who undergoes training? Defining the target group

If training is not designed specifically for the group to be trained there is no guarantee that it will work. The chosen objectives of a particular training course are, thus, closely interrelated to, and dependent upon, who is undergoing the training. It follows that the organization trained must define the target group. This requires both an understanding of the structure of the organization and its attitude to WID. So as to provide clear objectives that relate to work practices, it is necessary to understand how gender relations have a bearing on such practices in the organization concerned. This requires a knowledge of the power and decision-making processes within the organization and an understanding of the procedures most requiring change. The persons responsible for the definition, development and implementation of the institution's policies, programmes and projects also have to be identified. So do the strategically important allies who can provide leverage in the acceptance of training.

Where does training start in an organization? Three alternative scenarios can be identified, with important implications in terms of the entry point. Is the request for training from a 'top-down' mandate, a 'bottom-up' movement from within the organization or from a nervous organization concerned that they are 'losing out'? Each has important implications in terms of the structure of support necessary not only to carry out training but also to ensure that it is taken seriously. The evidence shows that gender training is most effective not only when conducted on a sufficiently large scale but also when it occurs within a favourable policy environment. A clear directive from top management, the necessary resources, and appropriate systems of accountability provide such a context. Strong institutional commitment will only survive if senior management involve themselves in the training itself – to provide the sustained support necessary to institutionalize such training. For such reasons, training often begins with, or includes in the early stages, senior policy-making staff, followed by programme-level staff.

How much do staff need to know? This may depend on their level of seniority and work responsibility. While very senior managers often only need to be aware of the importance of the issues, it may be the middle-level staff who require the specific techniques. Therefore, within an organization different objectives may be required for the various target groups. In most organizations, but especially those structured hierarchically, it is not appropriate to mix levels, given the sensitive nature of training. Senior staff do not want to be seen to be wrong in front of lower-level staff. The latter in turn are often intimidated, limiting their participation in the presence of more senior staff. However, different experiences can considerably enrich training courses. Senior staff may only be concerned with policy formulation, and can benefit from the contribution of those with operational experience. Ultimately, much depends on personal characteristics. Often very senior or junior staff are not hostile. It is those in the middle of their careers, for whom the defence of the *status quo* is most important.

Once an organization decides the most appropriate level to start training, and identifies the objectives for different groups, several further decisions have to be made relating to the composition of the staff trained. What are the common features, the criteria that bind them in training? Do they all work on the same sectoral concerns? Is it appropriate to mix staff from different sectors? Similar questions must be asked in relation to the sex of participants.

Is it more appropriate to have both men and women together in training? Finally, what is the most appropriate size of group for training, and length of time for a course? While answers to all these questions are contextually specific and depend on the culture of the organization, a number of general rules apply across the board.

The extent to which it is appropriate to mix professionals working on different sectors of development depends on the purpose of training and the methodology used. Because gender planning provides participants with tools that can be used cross-sectorally, the training is designed for mixed-sex groups. One of the biggest causes of tension in training groups with mixed groups is different levels of awareness. Where some participants are already aware of the issues and want training in gender planning, they can become impatient while waiting for colleagues to 'catch up'. In selecting target groups it is important, therefore, to ensure that participants have a similar 'level' of awareness of issues. Furthermore, it should be noted that mixed-sex groups often have different dynamics from single-sex groups. The training of women, whether in single-sex or mixed groups, brings home only too clearly the importance of not conflating the personal and the professional. For there is no a priori reason why women should be more sympathetic than men to the issues. Indeed, precisely because they are often professional women who have 'made it', they may be particularly hostile to gender training.

The size of the group also depends on the training methodology. The experience of gender planning training has shown that twelve to fifteen is the ideal number. Because of its highly participatory methodology, in larger groups there is less space for the participation of the whole group. This allows less interested participants to take a back seat. In smaller groups the lack of diversity of experience can make training less interesting. However, this question also relates to the length of training. Practice has not borne out the rule that the more senior professionals are, the busier they are, and therefore, the shorter the training period should be. Commitment to spending time on training depends on how seriously the organization takes training, and who has asked participants to attend. Short gender training courses vary from an hour or two with the aim of sensitizing and creating awareness, to seven to ten days in order to provide detailed skills. It is certainly true that the more time allocated to training the greater the skilling that can be provided. However, it is important to be realistic about what is possible, and to be able to recognize that a critical mass can be reached in terms of conflicting commitments. While the average length of many gender training courses is one day, practice has shown the optimal length for senior staff to be three days.

When is training done? The development of an institutional strategy of training

Whether training is considered important depends on the institutional strategy in relation to WID. Those organizations with WID specialists and those that mainstream WID issues differ identifiably in their strategy towards training. Many organizations still prefer to see training as an outside job that they hire consultants to do in a 'top-down', one-off manner. However, as is well known, others have recognized the complex interrelationship between the different components of a training strategy. They have sought, therefore, to develop one in relationship to their own policy, programme and project cycles, as part of a long-term overall institutional strategy. In deciding whether training is a priority, a clear relationship exists between the institutional strategy adopted and the extent to which training is considered important. Organizations that have separate institutional WID structures, which emphasize gender specialists, tend not to prioritize training. Since their strategy is to 'own' the issue, and they know it well enough themselves already, training is often very low in the order of priorities, if considered at all. In direct contrast are those institutions that prioritize the integration and mainstreaming of gender planning. Here training is often the most important instrument to ensure that the institution increases the number of gender-aware generalists.

This distinction also has implications for the extent to which training is considered a product or a process. Organizations 'coerced' top-down, or through outside pressure, to adopt training without understanding its importance, are more likely to identify training as a product. Outside consultants control the entire operation, which is packaged and delivered with manuals. Organizations that consider training a high priority recognize that it is a slow, costly and time-consuming process. Training must be tailor-made for the needs of the particular organization, and target groups. One-off manuals are less likely to be developed, with training tools developing as training itself gradually integrates into the institution.

When training is identified as a process, monitoring of both the content of the training and the identification of target groups is undertaken. Gender planning training at ODA, for example, began with one-day workshops with mid-level professional staff at headquarters in London. In the second stage it included regional development divisions in Africa, Asia and the Pacific, as well as the training, at their request, of more junior headquarter staff. In SIDA, the target group included headquarters staff, country DCs and WID officers, but in the second phase senior consultants participated. Training as a process can become a means to institutionalize gender policy. In both SIDA and ODA, for example, the introduction of a long-term training strategy has had important consequences in terms of changing planning procedures within the organization. Training has been the vehicle for the introduction of new gender policy which integrates such tools as gender needs assessment and the WID/GAD policy matrix. Both organizations recognize that without the introduction of these tools in training sessions there would not have been sufficient consensus for their adoption.

Who provides the training? Operationalizing training within an institution

To undertake gender training requires very specific skills, which are often not recognized. Two issues are important. First, training is not teaching. Just as the distinction between academic research on gender issues and gender planning is important, so the difference between teaching and training is equally important. The same pedagogic approaches are not necessarily appropriate. Training can be defined as the expansion or consolidation of technical skills to put knowledge into practice. It is not lecturing, with its emphasis on the transmission of theoretical knowledge and ideas. If teaching concerns the creating of an awareness, or consciousness, of the complexity of gender issues through an increased analytical capacity, training concerns developing the skill capacity to translate such awareness into very specific tools that can be used in practice. Training relies on a diversity of pedagogic techniques, such as exercises, small-group discussions, brainstorm sessions and so on. All are designed both to reinforce the skilling provided and to ensure that the participants themselves work through the issues so that they emerge confident to use them. In reality, training may require some teaching, but teaching can very successfully be done without training.

The second issue is that teachers are not necessarily trainers and gender planners are often neither. Frequently, in identifying professionals to undertake training in gender planning the assumption is that a gender planner, or an academic with a knowledge of gender issues, can train. As many problematic experiences have shown only too clearly, this is an incorrect assumption. Often trainers find it much easier to learn about gender planning, than gender planners or academics find it to learn how to train. The capacity of the trainer is one of the most crucial determinants of successful training. This therefore poses an important constraint that requires recognition at the outset in developing a training strategy.

In institutionalizing gender planning training, the critical question concerns the extent to which outside consultants on their own can successfully undertake training. This depends primarily on the purpose of the training. If the purpose is sensitization or creating gender awareness, an outside consultant can often be adequate. However, if the purpose of training is to translate skills into the specific contexts in which trainees are working, then senior co-trainers from within the institution are an essential prerequisite for success. In view of the time, expense and widespread reluctance of many institutions to accept this training principle, it is necessary to explain the reasons for its importance.

First, only those working within an organization ultimately know how it works. This input is essential if gender planning is to be operationalized in the work practices of those involved in training. Secondly, institutionalizing gender planning requires co-trainers who can articulate and identify the constraints and opportunities for this process, and provide the essential follow-up that results from successful training. While outside trainers may feel confident, they lack the necessary status and position. They have little impact when they identify bureaucratic blockages or suggest intra-organizational alliances likely to push the agenda. Ultimately, trainers working side by side from within and outside the organization develop a sophisticated understanding of the issues that are 'taboo' for each. There are some issues that it is more helpful for outsiders to raise, while there are other issues that only staff trainers from within the organization can bring up. Hannan-Andersson (1991), for example, has identified that in the SIDA training strategy two different types of input – from the consultants and from SIDA's Gender Office – were both complementary and essential. Consultants could provide a good background in gender analysis and gender planning. However, they could not help SIDA personnel and consultants in making the direct link to their own activities. Gender Office staff were vital to ensure this part of the process.

It has been noted that when training occurs without co-trainers, belligerent trainees often express aggressive hostility, or simply an off-hand dismissal of the WID institutional structure. Since it is difficult for the outside consultant to defend the WID office, this only serves to reduce the perceived lack of importance of gender planning. When co-trainers are present, the training enhances the importance of the WID staff and helps them in furthering their role as catalysts within the institution. Ultimately, training will only be taken seriously within an organization if they 'own' it. Nothing better helps the integration of training than the long-term visible participation of the senior WID/GAD staff as co-trainers. Much of the challenge to change existing practices occurs in the last part of the workshop, when excited participants ask, 'What can we do?' The capacity to hand over to the WID personnel who then take the suggestions further reinforces the issue of training far beyond the classroom. Because of gender planning training at ODA, for example, this type of process resulted in requests that the guidelines for project checklists should be rewritten.

A third reason relates to the status of training within the organization.

Many institutions identify training as a low-status, purely 'technical' activity, undertaken by junior staff with small budgets. This means that they view gender training in the same way. When senior staff became co-trainers, the status of training itself is 'upgraded'. Therefore many institutions when embarking on WID training do not initially involve the existing low-status training division. However, since the ideal training strategy is ultimately to hand over training to the organization, the training of trainers must be an integral part of the training strategy. Once there are sufficient gender trainers within the organization – whether they do it as one of several activities, or are hired full-time to train – the training becomes institutionalized into the ongoing mainstream training programme. While organizations such as USAID have a full-time training staff, others, such as SIDA and ODA, still prefer the strategy of using senior WID staff as co-trainers. In gender planning training two trainers work as a team, balancing active and passive roles in a highly structured manner. Not only is this type of training very stressful for trainers to cope with singly, but also the balance between the two provides an important dynamic. It can be helpful to have a male trainer. However, the capacity to train and the seniority of trainers are a trainer's most important assets, given the need to be treated with respect by senior staff.

How to train? The content of training

Obviously, the training methodology is the most critical determinant of the content of training. This then decides more specific details relating to such issues as the programming and format of the workshop, the structure of the content of each session and the materials required for each session. The Appendix outlines the detailed content of gender planning training. This general description of training strategies, therefore, concludes by highlighting the debate about the content of training. This concerns the extent to which it can be pre-packaged or whether it needs to be developed for specific contexts by trainers in collaboration with the organization concerned. To a large extent the objectives of training decide the answer. If the concern is to create awareness or increase ability in gender analysis, then training packages can be formalized for widespread use. If the objective of training is the translation of skills into planning practice then the working procedures of the organization must be understood. Training must be designed to meet the needs of different organizational groups.

At a specific level, the debate over course materials, particularly case studies, illustrates this difference. Are pre-developed case studies appropriate, or is it necessary to develop specific materials to train participants how to operationalize new procedures? Institutions training with the Harvard gender analysis approach base their training on pre-designed case studies.

FAO, for example, have developed eleven case studies covering a range of countries and technical areas and have used these, with gender analysis framework tools, to train some 1,000 professional staff (Howard-Borjas *et al.* 1991).

In contrast, given the particular focus of the gender planning methodology, it emphasizes the importance of using materials that are currently in use within organizations. It identifies this as critical since it makes the training central to the current work experience. Participants cannot, therefore, dismiss it as irrelevant to the institution concerned. It does, however, also mean that materials need to be more 'custom-built', and designed for specific training groups. Commonly these are project- and programme-level documents. SIDA, in its current training course, uses case studies based on SIDA documentation. These are not summaries of documents or versions especially written for training purposes. Case examples cover Asia and Africa, and review areas, such as water supply, road construction, hydropower and health (Hannan-Andersson 1991).

How to evaluate training? Evaluation procedures

To date there have been few systematic attempts to evaluate gender training quantitatively, other than questionnaires that assess the 'happiness quotient'. This is the level of satisfaction with training, but not the impact of training on work practices. Nevertheless, organizations do monitor the impact of training in a number of ways. Gender planning training provides participants with several tools that allow practitioners from different disciplines, and of different persuasions and positions on WID/GAD issues, to communicate in a common language. This can be monitored in terms of the resulting co-operation and co-ordination in work practice between both departments and individuals. However, one problem, identified by Christian Aid, is the distinction between taking on the rhetorical 'spirit of the language' and implementing it in practice.[6]

Anderson (1991) has recently outlined a potential evaluation approach to training through an input/output model that can reflect the variety of approaches taken to gender training. Inputs include the trainers, their materials, the content of training and the training method, the time, the place and the group who are trainees. She argues that when these elements work together, the outcome of training can be greater than the sum of the parts, 'a kind of magic can occur through which everyone gains a great deal' (Anderson 1991: 5). In contrast, when the elements do not come together, participants can learn very little, despite well-prepared training and effective trainers. Those who have undertaken training know only too well how logistical problems of timing or space, beyond the trainer's control, can only too easily result in a disastrous workshop.

Anderson (1991) provides a very useful threefold categorization of the outputs of training, as participant changes, institutional changes and external impact. She claims that the impact of training on participants can be identified in terms of changed attitudes, increased skills and knowledge and changed behaviour. The latter are the most difficult to assess. Institutional changes are easier to evaluate since they relate to changes in both institutional structures and operational procedures. These can result as much from 'bottom-up' pressure after training as from 'top-down directives'. Anderson does, however, caution that training may be only one of many initiatives to change practices. Then its purpose is to 'speed up' the restructuring.

The final output, that of the external impact, is the most difficult to assess, because of its distance from the actual training. Different objectives of training and their underlying rationales, as outlined at the beginning of the chapter, are obviously critical in the evaluation of training. Anderson distinguishes two external impacts of 'improved development' and 'increased equality'. The former can be used to evaluate training based on what she calls the 'efficiency hypothesis' (with an underlying efficiency approach to WID). Its major objective is to emphasize the importance of gender analysis as a tool for deciding how to use development resources more effectively to achieve goals in economic output. The latter can be used to evaluate approaches based on the 'equity hypothesis'. This focuses on gender in terms of its influence on the allocation of social and economic benefits from development, and thus, on equality between men and women. Obviously, any evaluation of the effectiveness of training ultimately depends on the approach taken. This suggests that ultimately, like training itself, evaluation indicators need to be designed to fit the particular approach of the organization concerned.

Concluding comments

Anderson's model of evaluation has yet to be tested. Nevertheless, from nearly a decade of experience the qualitative evidence clearly shows that institutionalized gender training within an organization plays a deciding role in a slow *process* of change. The objectives of gender training are determined both by the underlying approach to gender issues – women's release from subordination or more efficient development – as well as by its skilling objectives, to provide analytical skills or the capacity to translate into planning practice. Over time the importance of gender planning training has shifted from changing attitudes to increasing skills and knowledge. Training has been particularly effective in providing sympathetic gender-aware planners with the necessary tools and techniques that they require. It also provides a common language allowing for non-threatening communication.

In some organizations gender training has proved to be not an end in itself but a means to change. It has provided the catalyst for changes in both institutional structures and operational procedures. As participants themselves have confronted the ineffectiveness and inefficiency of palliative procedures and institutional structures designed merely to pay lip-service to WID issues, it may be useful to assess if 'bottom-up' demand for institutional and operational change is more successful than changes resulting from top-down directives.

With hindsight, gender training may prove to have been one of the most effective tools of the post-Nairobi decade to shift the WID/GAD agenda. Nevertheless, gender training, on its own, cannot solve the problem of women's subordination. There is a widespread preference for the efficiency of the WID approach (with its training objective to 'improve development'), rather than an empowerment approach (with its training objective of 'increased equality'). This indicates the vested interests at all levels. Ultimately, challenging such power can only be achieved through political intervention. In the last analysis gender planning also requires a political agenda – the subject of the concluding chapter of this book.

9   Towards an emancipation approach

The political agenda of women's organizations

Essentially this is a book about planning and planning procedures. Nevertheless, underlying the detailed description of gender planning, its methodology
and practice have been two important themes. First, gender planning is both
political and technical in nature; as with all other planning traditions, the
context in which it is situated determines its content. The differentiation
between practical and strategic needs provides the most important principle
of gender planning, allowing for the distinction between two sets of planning
needs. The planning methodology to meet practical gender needs is primarily
'technical' in nature, requiring the necessary tools and techniques to assist
women to do better what they are already doing. In contrast, the planning
methodology required to meet strategic gender needs is clearly 'political' in
nature. Secondly, as a transformative planning tradition, gender planning
assumes conflict in the planning process.

Gender planning is not an end in itself but a means by which women,
through a process of empowerment, can emancipate themselves. I argue that
this is best achieved through a process of negotiated debate about the
redistribution of power and resources within the household, civil society and
the state. Obviously in such a debate participation of women, gendered
organizations and planners is essential. This concluding chapter places
gender planning within its wider context, by bringing together these themes.
This requires a dramatic broadening of focus. It is important to stress,
however, that the identification of constraints and opportunities for gender
planning, in terms of both structure and agency, is itself an extensive topic.
Ultimately this will require further elaboration in another book.

Throughout this book, in examining the way that changes in the position
and status of women in developing countries have occurred, an important
distinction has been made. Change instigated through 'top-down' interventions of the state as the dominant 'structure' of power, control and domination
is distinct from change achieved through 'bottom-up' mobilization of
'agency' in civil society. Chapter 3 discussed some of the main issues relating

to the contents of 'top-down' state planning interventions as they relate to gender issues. It described the ambiguous manner in which the state, in particular contexts, both provides for, and controls, women, either directly through legal means, or indirectly through its control over institutional structures and planning procedures. This has shown only too clearly the limitations of both its political will and its ability to confront fundamental issues of women's subordination.

The development of gender planning comes out of the powerful social and political movements that women themselves now generate, rather than because of state intervention. As a new planning tradition, necessary to respond to this call for emancipation, gender planning is simply the engine to operationalize this specific political concern so that it becomes accepted as institutionalized practice, in what is an emancipation approach. As a transformative planning tradition, gender planning recognizes that emancipation can only be achieved by identifying contextually specific entry points. This defines where there is 'room for manoeuvre' for debate and negotiation, at the conjuncture of the state and civil society. In challenging rather than ignoring or avoiding conflict and tension, gender planning seeks to identify such politically appropriate points of entry.

The fact that planning can no longer be neatly compartmentalized into separate, isolated, 'top-down' or 'bottom-up' interventions means that criteria governing who is involved in the planning process, the relations between those involved, and the judgement as to what is an acceptable decision all become critical considerations (see Chapter 5). If negotiated debate is ultimately the most important way to ensure the implementation of gender policy, will 'voice', and the confrontation of those who hold power, at the level of ideology as well as materially, achieve it? Or is the use of 'exit', withdrawal and the creation of alternative planning interventions, a more appropriate strategy? There is no easy, one-off solution to this problem. Ultimately it will depend on the costs against expected returns whether 'voice' or 'exit' is the more appropriate solution.

If the success of gender planning depends on the participation of women, then it is the organization of women within civil society that requires examination. Because of their capacity to reach the 'grass-roots' where 'real people' are, organizations outside government, referred to throughout this book as Non-Governmental Organizations (NGOs), have increasingly been identified as the institutional solution for 'alternative' development models. This is equally true of women's NGOs. However, this has also resulted in a reluctance to examine critically their strengths and weaknesses. Consequently it is often difficult to identify their limitations and potential in challenging existing structures or in providing alternative institutional structures for planning.

This final chapter examines current Third World women's organizations and movements. It identifies their potential capacity to defend their own agendas through 'exit', setting up alternative organizational structures that can carry out change themselves. It also examines the potential of such organizations to influence change though 'voice', developing strategies for negotiating successfully, with those currently controlling resources and power, to ensure that they implement change. What are the choices that women make, both individually, as well as in groups? Where does the family end and the public domain begin? This chapter describes the mixed experience of women's organizations in raising consciousness to confront women's subordination, creating alliances and linkages to ensure the success of planning processes. In so doing it appraises DAWN's categorization of women's organizations. In addition, it highlights entry points identified by women's NGOs for negotiation and debate around women's needs at four different levels: household; civil society; the state; and the global system.

NGOs AS ALTERNATIVE ORGANIZATIONAL STRUCTURES FOR DEVELOPMENT? ISSUES OF DEFINITION

It is important to recognize that developments among Third World women's organizations in the past decade have not occurred in isolation. Although many experiences of such organizations are unique because they are specific to women, nevertheless they have occurred within a wider context in which the role of NGOs in the Third World generally has undergone fundamental changes. It is useful, therefore, to start by mentioning very briefly the relevance of some pertinent NGO debates to women's NGOs.

The 1980s saw the explosive growth in the number of organizations outside government. It has been argued that, if the 1960s was the development decade of growth with 'trickle-down', and the 1970s the decade of Basic Needs, the 1980s was the decade of the NGOs (Fowler 1988). This has certainly been true for women's NGOs. For example, a 1984 estimate identified 15,000 women's groups in Kenya alone, incorporating about 10 percent of the country's adult women (McCormack *et al.* 1986). In this context, the proliferation of women's NGOs seems no more than part of a broader phenomenon that recognizes NGOs as important actors in the development process. Official endorsement of their role is now widespread from governments, donor agencies and international organizations alike. A typical view, provided by the Bruntland Report, identifies the critical function of NGOs to ensure sustainable development. It recommends that governments should establish or strengthen procedures for official consultation and more meaningful participation by NGOs (WCED 1987).

Yet the role of NGOs obviously varies widely, dependent on their differing objectives. While it is understood that NGOs are not homogeneous, the residual nature of the term makes both definitions and categorizations fraught with complications. The same holds true for women's NGOs. Ultimately, the problem relates to the fact that most NGOs fall into a number of different categories, while different categories often have overlapping characteristics. Moreover, NGOs can be distinguished by several criteria, of which probably the most common are *institutional location*, *organizational composition* and *activity content*.

A distinction in institutional location is usually made between local, national and international NGOs (Cernea 1988). In terms of organization composition, international NGOs are most frequently *donor organizations*. National NGOs are generally *service organizations*, comprising those who support or cater to the needs of the 'grass-roots'. Local NGOs also can be these, but most commonly are *membership organizations*, which comprise people at the 'grass-roots' (Esman and Uphoff 1984).[1] Finally, the range of activity content of different NGOs includes both specific sectors such as health and education, as well as broader cross-sectoral concerns such as the environment, emergency aid, research, advocacy or even development itself. It is obvious that few NGOs fall neatly into one category or the other. More commonly they cover several categories, especially as they may well change their activity content, and to a lesser extent their organizational composition.

In reality, Third World NGOs are not a new phenomenon. Many, even if known by other names, have existed longer than governments. New characteristics of recent NGO development, including women's NGOs, however, include expansion in the scale and pace of increase of numbers and membership. NGOs now exist in most areas of local and national public interest. Characteristics also include a broadening of their functions from traditional relief and welfare concerns to development-orientated and production-support activities, including areas previously regarded as the 'exclusive prerogative of government'. In addition, they include an increase in their complexity and sophistication. This relates both to their internal structures as well as to their 'militancy and mobilization capacity' (Cernea 1988: 2). NGOs have strengthened contacts with other NGOs, establishing linkage systems between them. Where before only individual NGOs existed, now co-ordinated networks, councils and federations at national and international levels occur. The proliferation of such institutional structures during the past decade makes it important to identify the underlying reasons why NGOs have assumed such importance, in terms of both numbers and scope.

The social and political development agendas of NGOs

The most important organizational characteristic identified in NGOs has been their capacity to incorporate voluntarism into an organized structure. This results in a potential 'to mobilize people into organized structures of voluntary group action for self-reliance and self-development' (Cernea 1988: 7). Traditionally, their role has been identified as an alternative entry point for local-level social development. Their comparative advantage over government lies in the ways they relate to beneficiaries, as well as their freedom in organizing themselves (Fowler 1988). If government's relationship with intended beneficiaries is one of control, for NGOs it is one of voluntarism. This means that NGOs have the potential to adopt an unambivalent position of support, mutual trust and equality of interest with the intended beneficiaries. For governments, an authoritarian stance and dominance is particularly prevalent in the 'economically and socially stressed' Third World. It is this control orientation that has resulted in hierarchically structured bureaucracies. Fowler (1988) maintains that these are suitable for the maintenance of a stable state in a stable environment. However, their incompatibility with the needs of Third World micro-development results in structural inconsistencies. The identification of voluntarism as an essential characteristic of NGOs raises questions concerning the objectives of so-called 'community participation' within such organizations, and the extent to which it empowers women and men in the community (see Chapter 5).

In order to examine the potential of women's NGOs as alternative institutional structures in the planning process, it is necessary to mention the conventional wisdom concerning the comparative advantage of NGOs over government. Economic, organizational and political factors highlight this. *Economically*, NGOs with their component of voluntary unpaid labour can allow for greater cost-sharing. Since the early community development programmes of the 1950s and 1960s, carried out in Africa and Asia by post-colonial governments with the assistance of local NGOs, this has been widely critiqued as 'development on the cheap' (Mayo 1975). Later community participation projects of the 1970s and 1980s became more politicized in their concern to ensure participation in decision-making as well as in implementation. However, this fundamental critique remains. As discussed in Chapters 3 and 4, economic reform policies have resulted in cut-backs in state service delivery. This has increased the neo-welfarist role of NGOs as alternative service delivery mechanisms, relying heavily on unpaid labour, particularly of women.

Organizationally, because of the different ways they relate to intended beneficiaries, NGOs have the comparative advantage of more efficient and effective structures for local-level service delivery. Recognition that NGOs play a key role representing community interests relative to the state has resulted in a variety of attempts at decentralization. This includes the delegation to semi-autonomous or para-statal agencies, devolution to local government and the transfer of functions from public to non-government institutions or joint exercise of functions (Rondinelli 1981). Economic events in the past decade make it particularly important to raise such issues, as buzz words such as 'democracy', 'decentralization' and 'participation' again enter the planning agenda.

Decentralization is also linked to increased democratization, and the political processes through which local communities can more effectively challenge the state to meet their needs (Diamond 1989). In a recent assessment of NGOs, Cernea (1988) identifies several causes for the 'genesis' of Third World NGOs that go beyond the traditional domain of local-level economic and social development, to include more overtly political dimensions.[2] This shift in NGO agendas relates to changing agendas, as the state reduces the level of public control. It includes a shift in service provision from the public to the private sector, and from central to local government.

The stress on appropriateness, decentralization, self-reliance, popular participation, sustainability, the building up of local institutions and getting rid of excessive and oppressive bureaucracies, all show a potential willingness to identify more appropriate managerial and institutional structures (Marsden 1990). Across very different contexts, the re-alignment of government interests, required by fiscal constraints, has institutional implications. Debates are taking place over cut-backs in public expenditure, the rationalization of service delivery and devolution to the 'private' and 'voluntary' sectors. This renegotiation of the boundary of state intervention has important implications for the NGO sector generally and women's organizations in particular.

Not only has the nature of government intervention changed. In many parts of the world the capacity and capability of organizational structures outside government have increased during the past decade, as NGOs have renegotiated their relationship with the state. The impetus for this comes from hands-on, 'bottom-up' practical experience, particularly of groups in society subordinated or exploited in the development process, whether on the grounds of class, ethnicity or gender. In repressive or strife-ridden contexts, groups deprived of both national and international official resources for social development have felt this most profoundly. They have identified managerial and institutional structures based on self-reliance, with decentralized modes of decision-making. These have highlighted the close interrelationship between social welfare, economic well-being and fundamental political and human rights (Moser 1992b).[3]

Clearly, the relationship between the state and Third World NGOs has become increasingly important over the past decade, for political reasons as much as for economic and organizational ones. Whether this has equally been the case for women's organizations is the next important consideration.

WOMEN'S ORGANIZATIONS

Why should women organize?

Why should women as a group organize separately, and what, if any, interests do they have in common that bind them as a group? As Whitehead has shown, it is necessary to recognize that 'women' do not, and cannot, form a homogeneous category. Social relations between the genders vary from one society to another. In addition, 'women experience significant variation in their situations in those wider areas of political, economic and social subordination and inequality which are not confined to the social relations of gender' (Whitehead 1984c: 6). These differences make it important to identify the basis for women's solidarity, collaboration and commonality of interests.

Women are frequently divided on the basis of community, class, 'race' or religion. However, throughout history, in their gender-ascriptive roles they have informally organized together around a commonality of interests relating to their practical gender needs. An extensive literature documents the way in which women informally co-operate over such needs as child care, productive and reproductive labour, and credit. As the description on women and community managing in Chapter 2 illustrated, women's daily support systems for shelter, food, employment and protection of their environment all provide a critical basis for protest and action. This is true, although women generally have not 'formalized' it within an organizational structure.

External state influence has also been a critical determinant of the growth and success of women's groups. In Guayaquil, Ecuador, for example, a widely known procedure of petitioning services in return for votes by self-help committees exists. This provided a catalyst for initiating popular participation among newly settled communities. In formalizing this process an external donor, US President Kennedy's Alliance for Progress Programme, provided the greatest influence. In this programme USAID grant allocation was conditional on the setting up of *barrio* committees. The experience of local organization gained from this programme was also then important for the formation of later groups (Moser 1987b). In Kenya, Moore (1988) convincingly argues that contemporary women's groups owe far more to the workings of the modern state than to 'traditional' forms of mutual association and co-operation. One outstanding feature of the Kenyan state has been its mobilization of local populations through *harambee* ('let's unite') to participate in self-help development initiatives. Holmquist (1984) has identified this type of self-help as a hybrid phenomenon. It contains local initiative, leadership and decision-making as well as outside guidance, finance and even sometimes control.

Although practical gender needs can provide the basis for female solidarity, they also can divide women. This obviously occurs when women
prioritize different needs. However, it also can be based on kinship, class,
'race', ethnicity, age or patriarchy. Women's kinship networks, for instance,
are frequently as important for the maintenance of links between male-based
lineage groups as they are for women themselves. Particularly in contexts
with scarce or very unequally allocated resources, such as land, food or water,
antagonistic relations between different groups can often cause far greater
divisions between women. Finally, the isolation women experience when
secluded within their own homes, both because of 'cultural' norms and their
onerous work loads, often has made it difficult for them to organize, or to
sustain organization. The privatization of women's work in the domestic
sphere, and the dispersal of female kinship groups, therefore, have both
served to reduce the capacity of women for collective consciousness
(Whitehead 1984b). Constraints such as these mean that community-level
groups of women, such as those described in Ecuador and Kenya, typically
do not move beyond articulating the need to meet practical gender needs. Yet
women also share interests, as a gender, and it is these that relate to their
strategic gender needs. Here it is the recognition of subordination, whether
within society or the family, which provides the basis for co-operative
organization.

Historical documentation and oral histories of women themselves both
show that the origins of Third World women's organization around strategic
gender needs are not new. As described in Chapter 4, since the nineteenth
century Third World feminism has been an important force for change.
However, women have subsumed such struggles within nationalist and
patriotic struggles, working-class agitation and peasant movements, rather
than forming autonomous women's organizations. In the first half of the
twentieth century struggle focused on legal, political and educational rights.
Once achieved, most formal women's organizations in countries as diverse
as India (Caplan 1978, 1985), Kenya (Wipper 1975; Bujra 1986) and Colombia (Yudelman 1987), along with their First World sisters, essentially 'sold
out'. They turned themselves from reformist groups into social welfare
organizations, concerned with practical gender needs. With their privileged,
knowledgeable, middle- and upper-class membership, they reinforced the
tradition of 'taking responsibility for the plight of their less fortunate sisters'
(Hirschman 1984: 33). Lacking any consciousness of their own oppression,
they directed their attention at 'uplifting' lower-class women. Not only do
such organizations fail to address strategic gender needs. In addition, in
taking responsibility for social welfare provision, they comply and reinforce
the state's attitude to social welfare as 'women's concern'.

This description of problems associated with women's organizations could lead to the conclusion that their capacity to become a powerful social and political movement is severely limited. Individually, women's NGOs frequently appear weak, under-financed and undirected. Yet in total their changing agenda may represents an important movement outside government for changing women's lives. A number of important questions require consideration. If women's organizations traditionally have been divided, co-opted and weak, why then did the opening of 'political space' provided by the UN Decade for Women (1976–85) result not only in a world-wide proliferation of small-scale groups operating outside government, but also in a new level of debate about their role in confronting government? Why is it that their policy agendas increasingly have included not only welfare and well-being but also improvement in the status and rights of women in society?[4] Do women's NGOs have additional capacity beyond those of NGOs generally in terms of specific characteristics that unite or divide women? Is the current popularity of women's organizations simply a further extension of cheap labour and state control disguised under the media hype that currently surrounds NGOs generally? Do women's NGOs have the capacity not only to raise consciousness but also to engage in the planning process?

To answer questions such as these is critical, but also problematic, given the widespread hesitancy to evaluate NGO capacity. In attempting to do so it is necessary to address three important issues: first, to 'unpack', categorize and disaggregate with greater rigour such organizations; second, to identify with greater candour the constraints that limit women's NGO capacity; and third, to identify with greater clarity the entry points for alliances and coalitions of women's NGOs to influence the planning process. The remainder of this chapter addresses these issues.

How do women organize? DAWN's classification of women's NGOs[5]

Women's NGOs, as much as NGOs generally, are not homogeneous, with their classification equally problematic. Most frequently women's NGOs are also distinguished in terms of their *institutional location*, *organizational composition* or *activity content*. Although this can usefully describe economic and organizational characteristics, it does not, however, distinguish the policy objectives of different organizations and their underlying political agendas. To date there have been few substantial classifications of women's NGOs, other than that undertaken by DAWN. This is done,

> not from the usual viewpoint of donor agencies that wish to know which groups are the most suited to receive funding. It is done rather from the

broad perspective of building and strengthening our own movements and networks, that is *from the perspective of empowerment.*

(1985: 83; emphasis added)

As stated above, any classification is determined by its objectives. DAWN identifies the empowerment of women through organization as the key to achieving the goals of feminism – in other words, the emancipation of women. On this basis they identify six different types of women's organizations. These vary in their aims, funding and methods of operation, and their potential capacity to assist in the empowerment of women. The organizations identified fall into overlapping categories: *outside initiated* and *small grassroots* organizations identified in terms of their institutional location; *worker-based* and those *affiliated to a political party* identified in terms of organizational composition; finally, *service-orientated* and *research type* identified in terms of their activity content. To this list DAWN add a seventh, a coalition of organizations around specific issues.

Table 9.1 identifies different organizations and their dominant policy approach. *Service-orientated* organizations are the first type. These most frequently have a welfarist approach. Their objective is to augment the role of social welfare agencies by meeting the practical gender needs of poor women relating to health, education and other services. They are essentially top-down, middle-class voluntary organizations with a low level of participation by local women limiting themselves to the objective of cost-sharing. As outlined in Chapter 4, historically, service-orientated NGOs have been closely linked to the modernization model of development. Now, however, they also include the neo-welfarist delivery models reappearing under recession and macro-economic models of structural adjustment. Such organizations take different forms, including middle-class women's clubs, such as Nigerian clubs (Enabulele 1985); those with an international affiliation, such as the YWCA, Women's Councils, Girl Guides and Ladies' Leagues (Lapido 1981; Lee 1985); and those originating from social reform movements or nationalist struggles (Jayawardena 1986).

As DAWN identifies, their weaknesses include the fact that they are elitist in bias. They employ top-down decision-making processes in which there is little scope for empowerment. In addition, they have no clear theoretical understanding of gender subordination, or its links to other forms of oppression. Nevertheless, such organizations are important for several reasons. Because of their membership they often have both direct and indirect links to policy-makers. They command significant resources, and have systematic methods for transferring skills and building leadership (DAWN 1985). For many feminists, such organizations are anathema (as mentioned above) because of their role in the reproduction of class relations. However, given

Table 9.1 Characteristics of DAWN's classification of women's organizations

DAWN's classification of women's organizations	Service-oriented	Affiliated to political party	Worker-based	Outside-orientated	Grass-roots	Research	Coalition of organizations
Predominant policy approach	Welfare	Equity? Welfare Anti-poverty	Equity Welfare Empowerment	Anti-poverty	Empowerment	Equity/Empowerment	Empowerment
Level of interaction	Local	National	National/Local	Local	Local	Local/National	Local/National/International
Object of participation of local women	Low – for cost-sharing	Low – for cost-sharing	High – for efficiency, capacity building	High – for cost-sharing and efficiency	High – for capacity building and empowerment	Low to High – for capacity building	High – for empowerment
Opportunities	Meets PGNs through delivery of welfare packages; offers organizational experience for women	Potential to raise SGNs at highest level	All women workers' associations; seek to empower women but have weak resource position	Persist as long as generate outside funds	Specificity and relevance in meeting PGN as means to address SGN	Research useful for other organizations involved in action	High mobilization capacity
Constraints	Elitist class bias: top-down delivery; gender-blind	Danger of co-option with SGN subsumed to wider political goals	Former man-led unions prioritize male production concerns, not the PGN of women workers	Lack of theoretical analysis of subordination; among the weakest organizations	Inadequate resource base; middle-class urban bias	Danger of shifting to a less participatory top-down equity approach	Lack of a clear organizational structure; issue-based so lack of ongoing commitment

PGNs = Practical Gender Needs
SGNs = Strategic Gender Needs

their significant power, in specific contexts, it may be constructive to learn from their ability to raise women's issues in the public arena. This includes the identification of appropriate strategies to co-opt them for specific strategic gender needs. Bonepath (1982), for instance, argues that in the United States, older-established groups are often willing and motivated to co-operate with superficially more 'radical', newer groups. Such groups put contacts, expertise and financial resources at the disposal of more radical agendas.

Secondly, are those organizations *affiliated to political parties*? In principle they have an equity approach, although in practice their policy approach is more likely to be either welfare or anti-poverty. If they achieve equity, the approach is top-down and does not involve the participation of women. However, such organizations may include the participation of women for the purposes of cost-sharing and efficiency. The structural position of these organizations is both their strength and weakness. By virtue of being within a party they potentially have space to raise women's issues at the highest level. However, their autonomy to raise and address gender issues is questionable. There is a constant danger of co-option, with strategic gender needs subsumed to wider political goals. Very few political parties are explicitly feminist in orientation or internal structure. The co-option of female participation in order to further the wider aims of organizations that do not accord any genuine political status to women's struggle is a widespread phenomenon. It occurs in a diversity of politically aligned organizations. As mentioned in Chapter 3, nationalist struggles often highlight the problem. Here women's mobilization around gender needs, particularly strategic ones, may conflict with the more general aims of political struggle. Organizations such as AMLAE in the Sandinista period in Nicaragua, and OMM in Mozambique appear to represent the forefront of radical change. However, in reality they lack any genuine autonomy to define an agenda outside that of the wider political concerns of their parent organization (Molyneux 1985a; Peake 1991).

Worker-based organizations are a third type. These include trade unions of formal sector workers, as well as the organizations of low-income, self-employed women, such as market sellers. DAWN makes a critical distinction between the two. In trade unions mainly male leadership makes decisions, while women at the base participate in carrying out policy. In contrast, women's representation is much stronger in self-employed associations. The policy approach of worker-based organizations, in principle, is equity. In practice, formal sector unions prioritize production concerns. They rarely consider as a priority the reproductive concerns of women workers such as child care, maternity leave and associated benefits. In contrast, all women workers' organizations, such as SEWA, whether or not they have an underlying empowerment approach or are explicitly aware of women's subordination, 'tend to be very successful in empowering poor women in their own personal life situations' (DAWN 1985: 84). One important constraint of such organizations is their weak resource position, reflecting the poverty of their membership.

Organizations that have mushroomed during the Decade, because of external funds, are the fourth type. DAWN identifies most *outside-initiated* organizations, with an underlying anti-poverty approach, as concerned with small income-generating projects. Primarily top-down in approach, the participation of local women is greatest at the implementation stage for effectiveness and cost-sharing. Both the funding source and the approach means that many such organizations lack a theoretical analysis of women's subordination and its links with broader economic, social and political issues. DAWN assesses them as among the weakest of the different types of organizations. With no previous organic history, and little independent organizational or resource base, they persist only as long as they generate outside funding (DAWN 1985).

Grass-roots organizations are the fifth type. These originate from the economic and material conditions women experience, and with their underlying policy approach, frequently empowerment. Such organizations focus on meeting practical gender needs relating not only to income but also to health and education. These are a means by which they raise consciousness to address strategic gender needs such as domestic violence, legal rights and political struggles. DAWN identifies that some of these groups are explicitly feminist in orientation. Their main weaknesses include an inadequate resource base and overwhelmingly middle-class, urban membership. Nevertheless, their focus on local- and household-level issues means that such organizations have a greater potential of empowering their membership than many other types of groups (DAWN 1985).

The sixth type of organization are *research organizations*. These focus on participatory action research, women's studies associations and women's networks. When such groups are 'committed to using their findings to serve and empower the subjects of the research' (DAWN 1985: 85), they have an empowerment approach. Their research provides useful background to inform and back up political action by other organizations. However, the considerable potential that such groups have to influence public policy debate means that they can easily shift towards a less participatory equity approach. Research organizations need to develop structures and methods of accountability to both action organizations and the subjects of the research. Only then can they avoid the development of rifts between themselves and other groups (DAWN 1985).

Besides these six types of organizations, DAWN identifies numerous women's movements. These encompass individuals, organizations and coalitions that have sprung up during the Decade around specific issues such as racism, political repression or sexual exploitation. As with issue-based organizations, their strength lies in their mobilization capacity, their weakness in a lack of clear organizational structure and ongoing commitment.

Time, change and convergence within and between types of organizations

On the face of it the evidence shows that some types of organizations have greater capacity than others to move from practical to strategic gender needs. Obviously, however, it is not enough simply to classify women's NGOs. Although DAWN's classification provides a useful starting point for appraising the capacity of different types of organizations to empower women, nevertheless it is limited in several significant respects. In the seven-year interval since first produced, important external political and economic changes have occurred. These have included the fall of the Berlin Wall and the break-up of the Soviet Union, the widespread adoption of economic structural adjustment measures throughout the Third World, and the continued resurgence of nationalism and religious fundamentalism. Such events highlight the complexity of confronting and challenging the power of the forces of subordination at global, regional, national and local level.

Equally during this period women's NGOs have not remained static, as the analysis so far may suggest. Because of its wide usage, I have retained this sixfold categorization. However, convergence has become more explicit during the past five years. As resources have decreased, organizational processes of consolidation or fragmentation highlight underlying objectives with greater clarity. For instance, a tendency for convergence between worker-based and grass-roots groups can be observed, as well as a greater clarification of approach of outside-initiated groups. A diversity of internal and external constraints and opportunities that cut across organizational types determine the capacity of organizations to influence the planning process. While DAWN highlighted some, others have become more significant in the intervening period. In addition, organizations have begun to recognize that women's empowerment within the confines of a project does not automatically lead to their emancipation in the planning process. This has resulted in a shift in emphasis towards coalitions and networks. These require greater compromise and more realistic understanding of both political and planning realities.

Empowerment and emancipation ultimately depend on multi-faceted intervention strategies, in which different types of organizations act at times individually, but also at other times collaboratively. To identify when, where and why politically appropriate entry points occur to shift the agenda requires the introduction of a more dynamic analysis. There is no one solution. In some contexts clearly defined 'voice' strategies can be successful to confront the issue. In other contexts the use of 'exit' and withdrawal as an alternative method of protest can be more appropriate. In the world of political power there are no easy technical rules and regulations. Nevertheless, general guidelines can be helpful. The final section of this chapter attempts to provide these. It highlights some of the main internal and external constraints and opportunities for women's NGOs to empower and emancipate themselves through the planning process.

THE INTERNAL AND EXTERNAL CONSTRAINTS AND OPPORTUNITIES OF WOMEN'S NGOs

One significant difference between many women's NGOs and other organizations relates to internal organizational and leadership structure. This presents comparative advantages and disadvantages. DAWN, for example, cites the search for non-hierarchical and non-formal organizational structures as problematical in an increasingly formalized and hierarchical world. It prevents the establishment of clearly delineated relationships with complex and bureaucratic decision-making bodies, necessary to pressure them to carry out policies in the interest of women's NGOs. Such groups have a historical mistrust of the manner in which centralized power has been used as an instrument of subordination. This has led to innovative ways of sharing responsibility that do not reinforce existing relationships of domination. However, unstructured organization and unclear authority demarcation and defined responsibilities also can lead to a lack of internal commitment. This occurs especially among volunteers, and results in conflict, power struggles, and weakens 'voice' (DAWN 1985).

Managerial and organizational problems also become particularly problematical as organizations expand and grow. With the shift from a small grass-roots body to a large organization, problems can easily arise. These concern when to set up new structures, where to place authority, and how to settle conflicts. Weak management structures are particularly open to inefficiency when women's NGOs expand their areas of commitment into inter-sectoral planning – reflecting women's different needs in health, services, water and transport. One obvious consequence of work overload is conflict and dissent.

Despite their non-hierarchical nature, women's NGOs frequently have strong leaders. In her evaluation of five women's NGOs, Yudelman (1987) highlights the importance of leadership in organizational success, providing a salutary exposé of the 'uses and abuses of charisma'. Strong leadership can be particularly important in getting NGOs started, when leaders inevitably have to deal with tenaciously held assumptions about women's roles. Thus, the ability to conceptualize and articulate women's problems and to organize people, engender loyalty and commitment, and resources, are all vital leadership qualities. The problems inherent in strong leadership, however, tend to arise over time when it can create dependency, resentment and internal conflict. If the organization becomes too closely identified with the leader it can have difficulties institutionalizing itself, with the leader in turn unwilling to share power or delegate responsibility.

The internal conflict between leadership and staff common to many women's organizations has been attributed to the strong emotional commitment such staff have to an organization. In addition, women's ambivalence towards the exercise of public power and their desire for more participatory, open organizations also can cause problems. Solutions to the problems posed by strong leadership partly lie in the establishment of tighter management systems. They also include training of a second generation of leadership ready to take over. Opening up the leadership, delegating responsibility and authority, and ensuring that staff have access to training for professional advancement all help this problem (Yudelman 1987). Internal conflict is probably the most divisive internal constraint influencing the sustainability of women's NGOs. Strong loyalties and sisterhood solidarity also makes it a highly sensitive issue to put on the planning agenda.

Perhaps the most important external constraint affecting women's NGOs is that of finding appropriate 'entry points'. For planning interventions to be effective, they must relate directly to the 'points of power in society'. Because women's NGOs are not homogeneous in terms of the level at which they work, 'entry points' for negotiation and debate will differ depending upon the focus of their work. Recognition of the level at which an organization works and its own aims is therefore essential before strategies to empower and emancipate women can be identified. DAWN identifies three different *spatial* levels of intervention and empowerment; first, the global or supranational level where organization for the common goals of a more just and equitable international order concern such issues as disarmament, debt crisis, sex tourism and the multinational sector; second, the regional and subregional level, particularly important in supporting women in countries that are politically repressive (such as South Africa), or in which the state has attacked women's social and economic position (such as some Islamic countries); and third, national-level mobilization relating to strategic gender needs such as laws and civil codes (DAWN 1985).

For the purposes of gender planning the determinant is not so much the spatial level as the *focus of power and control*. Therefore the subsequent analysis identifies the following four levels: the *family*, *civil society*, the *state* and the *global system*. The manner and extent to which entry points for negotiation and debate differ at each of these levels requires identification. Ultimately success depends on the capacity to identify the appropriate issue for debate at each level.

Negotiation and debate at the intra-household level: autonomous or mainstream groups?

In many societies the starting point for the subordination of women is the family, one of the most powerful and pervasive mechanisms of control. The entry point for debate and negotiation at the intra-household level for women can come at an individual level from increased economic independence. As Bhatty in her study of women workers in the *beedi* (crude cigarette) industry in Allahabad, India, states, 'A greater economic role for women definitely improves their status within the household. A majority of them have more money to spend, and even more importantly, have a greater say in the decisions to spend money' (1980: 41).

Moreover, the solidarity and support provided by local-level organizations can play a critical role in consciousness-raising and solidarity, as much as in opening economic opportunities and providing skill training and child care. As discussed in Chapter 2, Sen (1990) emphasizes the importance of women's extra-household organization for changing perceptions relating to women's perceived contributions of 'gainful' work, and their relative share to family entitlements. Local organizations also provide the entry point for community-level debate and action relating to more strategic intra-household issues such as domestic violence and fertility control. The establishment of rape crisis centres, battered women's homes and well women's clinics are all local planning solutions to help women have greater control of their bodies, and their lives.

One important debate relating to local community-level organization concerns the composition of such groups, and whether they should be mixed or segregated. Is there a need for separate women's organizations to avoid the marginalization so frequently experienced by women when integrated into mainstream groups? The disadvantages of mixed groups are well known, particularly the decrease in autonomy experienced by women. Male domination of the leadership, management and finance all tend to result in male priorities in terms of goals, policy and projects. This also results in a loss of funds for women's programmes, and a general pressure on women to conform to traditional roles with accompanying 'female' programmes of health and welfare.

Negotiation within civil society: class versus gender

Opportunities for negotiation and debate at the level of civil society are constrained by the conflicts and difference in interests and needs between women based on class, 'race', religion or ethnicity. For middle-class women running welfare-service delivery organizations, this can be more a consequence of their lack of comprehension of the problems experienced by poor women than their concern to preserve class interests. Working-class women preoccupied with practical gender needs are frequently unwilling or unable to support mobilizations or join organizations concerned with more strategic needs. As Caplan so aptly states, 'Philanthropy provides a cross-cutting tie between the classes which masks the fact that their interests are opposed' (1978: 125).

Women's organizations, therefore, can only negotiate on issues where there is sufficient unity of interests due to gender, despite other causes of division. Over time, social and economic processes occurring in society influence this, with class transformations themselves often deciding what women's organizations can achieve. The Fakalakalaka Women's Movement provides an interesting example. This grass-roots network of self-help, income-generating women's groups was started in Tonga in 1978 by local Catholic nuns. Initially, it was an immensely successful, rapidly growing movement. By 1984, it had 400 groups involved in saving schemes, home-improvement projects and community-level activities. Small (1989) shows that initially the groups provided women with a means to negotiate at the intra-household level. This provided access to household funds they would otherwise not control. 'Individual women simply would never have been able to siphon off household income individually for development or home improvement. They would have had to contend with the demands of the household and kin in the use of household income' (1989: 5). The gradual collapse of the movement was due to conflicts caused by growing class transformation in Tongan society. As better-off women wanted to spend group money on different things from poorer women, dissent grew over spending decisions and the latter began to drop out. At the same time, wealthier women, who could afford more, increased membership fees, which accelerated the drop-out of poorer women (Small 1989).

In contrast to this there are examples of organizations where women of different classes have effectively worked together. An example of an organization to empower low-income women is the Centro de Orientacion de la Mujer (COMO), a worker-based organization in Mexico. Middle-class, educated women in Ciudad Juarez initially set up COMO in 1968. Its purpose was to assist in tackling the problems of working-class women factory workers in the *maquiladoras* assembly plants. These included the provision of welfare services such as child care, better transport and counselling. By 1975 it had turned into a militant working-class women's movement, confronting plant management by pressing for reforms in working conditions. A highly participatory three-stage approach provided the basis for this shift; first, the increasing involvement of women workers in the organization's work, through participation in community schemes, health, literacy and nutrition courses; secondly, skills training and local employment of women through COMO diploma courses in nursing, teaching, social work, secretarial work and English language; and thirdly, the development of other employment creation schemes for laid-off women workers, such as worker-managed production co-operatives. In COMO, what began as a middle-class professional women's organization shifted to a working-class women's movement, via participation and the empowerment of poor women. Here the participation of middle-class women was seen as an advantage, not a hindrance. Dr Valdez, a woman social psychologist and COMO's leader, and other middle-class members, had contacts to several Mexican presidents, the press and academic institutions. They therefore had the necessary social position, connections and skills crucial to getting COMO off the ground (Yudelman 1987).

Negotiation and debate at the state level: political support or control?

Does the political or ideological orientation of a country determine the process of negotiation and debate for gender planning by women's NGOs? Jayawardena, in her historical analysis of women's organizations, writes that, 'Women's movements do not occur in a vacuum but correspond to, and to some extent are determined by, the wider social movements of which they form part' (1986: 10). Simultaneously she shows that women's movements in Asia integrated into national liberation struggles against colonialism. After gaining liberation, however, they neglected women's issues, allowing them to fall into the background. Consequently, she argues, there is a need for autonomous women's organizations with their own goals, to avoid submergence in nation state goals. The extent to which the wider political context is a constraint that determines NGO capacity to empower and emancipate women is highly complex. It varies depending on different ideological interpretations of development, and their interrelationship not only to patriarchy and class, but also to race and religion. This in turn has important implications in terms of the composition of women's organizations, as autonomous or integrated institutions.

Frequently, it is short-term coalitions on specific issues that provides the entry point for women's organizations to negotiate at the level of the state. One powerful interest to unite women, on the basis of gender, is the state's attitude, in theory and practice, with regard to the issue of rape. In India, a classic example occurred in 1972 when two policemen raped a fifteen-year-old tribal labourer, Mathura. The Supreme Court finally acquitted them on the grounds that since no wounds were found on her body, it could be presumed she had not resisted and thus consented to sexual intercourse. This is a typical example of a rape victim being found guilty. It provoked a massive response only after four university law lecturers (two men and two women) wrote an open letter throughout India protesting against the case. In 1980 the protest and action provoked women's groups in Bombay to form a coalition, the Forum against Rape. A vigorous, organized campaign around local rape cases followed. A National Conference against Oppression of Women then provided the basis for national-level debate on the rape issue. Coalition pressure to change the law led to a discussion in Parliament. In broadening its agenda, the coalition then changed its name to the Forum against the Oppression of Women (FAOW). Although this soon fractionalized, due to dissenting views, it nevertheless succeeded in getting the issue of rape onto the political and planning agenda. Many different organizations in the past decade have in turn reinforced this action (Omvedt 1986; Desai 1990).

Poignant but highly controversial coalitions between women can occur when they collaborate under conditions of conflict.[6] Women's complicity against other women in the context of repression and war is a highly sensitive issue, as the experience in Nazi Germany showed only too clearly. However, women also can be at the forefront of conflict resolution, when their power is in non-political action. The Israeli Women in Black Movement protesting at the *intifada*, Protestant and Catholic women's groups in Northern Ireland coming together over the rights of women prisoners, and black and white women in South Africa dealing with 'conflict creatively' provide only three examples of coalitions. Such coalitions form when women question their stereotyped roles as 'protectors of values and culture' and 'mothers of the nation'. They reject the ethics of 'preparing their sons to die for their countries', and demand to be 'part of the answer not the problem'. However, such coalitions are open to violence and abuse, particularly of a sexual nature with accusations of 'sleeping with the enemy'. The only defence women have is to recognize that 'they are good at not quitting' and that strength is in their 'steadfastness'.

Negotiation and debate at the global level: is the North–South divide irreconcilable?

The International Women's Decade provided the 'political space' for formal, 'top-down' coalitions at the supra-national level. A diversity of global networks, particularly within the UN and Commonwealth, remain an important forum for international debate on gender concerns. However, as described in Chapter 6, it has been an uphill struggle. The fact that even in 1992 not all countries have ratified the Forward Looking Strategies suggests that for many it remains essentially a matter of lip-service.

'Bottom-up' coalitions to negotiate and debate global gender concerns have been more tentative and cautious, but also probably more realistic. Most women in both the North and South are caught up in their own parochial realities. Women in the North only too easily think that women in the South have to catch up. Writing in 1989 I provocatively stated, 'It may be that women in the UK can learn much from their better organized sisters in the Third World, who long ago learnt the limitations of relying on the state to reduce their dependence on men' (Moser 1989c: 184).

Many activist, feminist women from the South are clear that there are more issues to divide them from their sisters in the North than there are to unite them. Not only have imperialism and colonialism left brutal scars. In addition, they identify the re-colonization of the Third World working through such instruments as GATT, and the loss of sovereignty and state legitimacy through structural adjustment programmes, as mechanisms that serve simply to reinforce the existing North–South divide.

Despite such difference, during the past decade, international and regional issue-based networks have proliferated. These encompass the full range of women's practical and strategic gender needs, from health to housing, and include institutions of every type, from trade unions to women's institutes. As described in Chapter 4, DAWN has been critical in identifying *global gender issues* for debate. These it identified in its analysis of the interlocking crises of debt, food, energy and deterioration of social services, environmental degradation, political conservatism and religious fundamentalism and militarization. This has been important in forging 'a close relationship between the work of the emerging international women's movement and the socio-economic and political processes of the decade' (Antrobus 1989: 75). However, structural adjustment programmes essentially remain gender-blind, as described in Chapters 2 and 4. This provides a salutary example of precisely how limited the impact of global networks has been to date, in shifting agendas relating to either practical or strategic gender needs in the context of economic reform. Negotiation and debate at the global level is the most difficult to accomplish, but the most important to achieve. As Peggy Antrobus of DAWN and WAND Barbados has said, 'it is time for sisters in the South and North to start listening and talking to each other'.

CONCLUDING COMMENT

The purpose of this book has been to describe the development of gender planning as a legitimate planning tradition in its own right, with its particular conceptual rationale, its planning process, and its WID/GAD principles, tools and methodology. In the implementation of practice the evidence clearly shows that appropriate operational procedures, institutional structures and training strategies all play a critically important part in ensuring the success of gender planning practice.

This book shows that gender planning is not an end in itself, but simply the means by which women, through a process of empowerment, can emancipate themselves. As the century comes to and end, women increasingly are negotiating and debating to redistribute power and resources within the household, civil society and the state. Hopefully, gender planning, preliminary though it still is, can provide assistance to them in this quest.

Appendix:
Gender planning training
Its methodology and content

This chapter outlines the gender planning training methodology. It discusses such issues as its objectives and approach, the programming and format of workshops, the structure of the content of each session, and the materials required.[1] The methodology presented here describes the basic training process developed up to 1990, and still widely used today. For more recent additional work, see Levy (1991) and Hannan-Andersson (1992).

THE OBJECTIVE OF GENDER PLANNING TRAINING

Chapter 5 defined the aims of gender planning as the emancipation of women and their release from subordination. It identified training as one of a number of interventions by which this objective could be achieved. Thus, in line with Anderson (1991), gender planning training has the potential of a powerful 'transformative tool' through which people can learn new attitudes and skills to influence practice.

Gender planning training has three specific objectives. These are similar to three of the objectives of training identified in Chapter 8: sensitization; skill transfer in gender analysis and diagnosis; and most importantly, the translation of skills into planning practice. Gender planning training is based on the principle that it is a gradual sequential process of 'block building', with the different objectives reflecting stages in this process. First, they provide participants with an understanding of the issues of gender, planning and development as they relate to the different sectoral planning concerns identified by the organization undertaking training. Secondly, they introduce participants to gender planning tools that they can use to appraise and evaluate the organization's policy, programmes and projects for specific sectors. Thirdly, they enable participants to integrate the gender planning methodology into the organization's operational planning cycle as well as into the institutional structure of the organization.

THE APPROACH TO TRAINING

The gender planning methodology described through the course of this book provides the basis of this training approach. This concerns both the conceptual framework and the methodological tools for incorporating gender into planning, within the context of development. The approach can be summarized as follows.

It starts by recognizing the difference between sex and gender, and the fact that relations between men and women are socially constructed. It differs from a 'women in development' (WID) approach which, while recognizing the critical role of women in the development process, does so without necessarily referring to the nature of women's subordination. An understanding of the social construction of gender allows for the recognition that because men and women play different roles in society, they often, consequently, have different needs. This provides the underlying conceptual rationale for the training. It asserts that in the identification of the extent to which needs are met in policies, programmes and projects, it is important to disaggregate within communities, households and families based on gender.

Gender planning questions current planning stereotypes. These tend to assume that the structure of low-income, Third World families is nuclear; that in the division of labour within the family the husband undertakes productive work, while the wife does reproductive work; and that within the household there is equal allocation of resources and power of decision-making. It identifies that women have a triple role as reproducers, producers and community managers, while men have a dual role in productive work and community politics. It highlights the problems experienced by women in balancing their triple role, especially those of women who head households. It also reveals the extent to which, within households, men and women unequally share resources. In the identification of needs this approach makes a critical conceptual and planning distinction between practical and strategic gender needs. This allows planners to distinguish between interventions that help women to perform more effectively and efficiently the activities they are already undertaking, as opposed to interventions that assist women to achieve greater equality, and thus, transform existing roles. It then uses various methodological tools to evaluate different WID approaches. This is done in terms of their different objectives, with a distinction between welfare, equity, anti-poverty, efficiency and empowerment approaches.

The basis of this approach is *planning*, rather than analysis, and *gender planning* rather than with planning for women in development. The purpose of the training, therefore, is to provide tools not only for analysis but also for translation into practice. The methodological tools identified simplify complex theoretical feminist concerns identified above, such that they can be translated into specific interventions in planning practice.

Tools such as gender roles identification, gender needs assessment, the WID/GAD matrix and gendered participatory planning procedures help planners in gender planning procedures such as gender diagnosis, gender objectives and gender monitoring. They assist in showing planners how to ensure gendered consultation and participation in the planning process, as well as the mechanisms to identify entry points for planning practice. These relate both to institutionalizing and operationalizing gender planning. These include the appraisal and evaluation of complex planning interventions, and in the formulation and implementation of more gendered policy, programme- or project-level interventions, within particular socio-economic and political contexts. The integration of methodological tools into planning practice emphasizes the interrelationship between 'technical' and 'political' constraints.

THE DESIGN OF GENDER PLANNING TRAINING COURSES

As a 'transformative' tool, gender planning training can only be effective when identified as a process. Consequently, the first task in undertaking such training is the development of a training strategy. The organization undergoing training undertakes this collaboratively with the trainers. (Chapter 8 described the components of a strategy.) Once they identify and agree the strategy, the identification of the participants and the design of the training course itself is the next activity.

Organizations themselves decide the composition and number of participants in collaboration with the trainers, as part of the training strategy. Training groups may be organized on either a peer group or a hierarchical basis. They may vary in size from ten to twenty, since the methodology is not really appropriate for larger groups. Two trainers, working as a team, undertake gender planning training. One trainer should be a staff member of the organization with a knowledge of its policy and practice on WID/GAD issues. He or she acts as a resource person during the training, as well as providing critical training inputs. In undertaking training initially, a consultant works with institutional co-trainers. However, the strategy includes rapidly moving towards institutionalizing training entirely within the organization concerned.

Gender planning training is structured in terms of sequential modules. Based on the logic of a block-building process, a number of clearly defined modules introduce participants to the gender planning methodology. Each module is a 'block' that increases the participant's capacity to analyse gender, development and planning issues in the countries and sectors in which they operate, and to apply the methodological tools for intervention in these contexts. In this training methodology there is a correlation between the number of modules built into a course and the potential skilling of participants. Simultaneously, the in-built time flexibility ensures that the methodology can be used to meet the diverse training requirements of different institutions.

The number of modules in a training course depends on the length of time available and the pace considered appropriate for training. A one-day course, for example, can be divided into three or four modules. It can contain a lecture, three exercises (one of which uses several projects of the organization) and a final workshop. Participants undertake lectures and the final workshop as a group. They do the exercises in small groups of three to four. Depending on the total size of the group, exercise report-back sessions are undertaken in one large group or two medium-sized groups. Components may vary considerably within each module, with different training techniques used to introduce and reinforce the conceptual rationale and its translation into practice.

Each training course has a common structure irrespective of the sectoral issue addressed. This reflects the training process involved in acquiring the necessary methodological skills to operationalize gender planning in the different planning procedures of the organization concerned. Substantive issues relating to the organization's sectoral and country concerns are identified, as is its planning methodology. They are illustrated through such techniques as lectures, films and exercises, the latter based on carefully selected case-study material and supplemented by small-group work, discussion and individual work.

Not only do organizations differ in terms of the time they allocate to training; in addition, the pace of training varies depending on the type of institution. As a generalization NGOs are more reflective and require longer to debate the issues raised, while donors are impatient to get to 'hands-on' practice. Ultimately, these two factors decide how many stages of the process participants have the time to absorb in a training course. It also means that trainers must be pragmatic, recognizing the time constraints imposed by the organization, and develop the skills to vary the scope and pace of the workshop.

Table A.1 (see page 228) provides the timetable for a prototype one-day gender training course.[2] Given the limited time and the fact that it includes four modules, a very fast pace is assumed. This timetable, therefore, is most appropriate for training middle- and senior-level staff in organizations that only allocate one day to training. Although the fourth module is critical, on the block-building principle it cannot be introduced before participants have been through earlier modules. In a three-day workshop, where a slower pace is possible, one-and-a-half days is generally allocated to this module alone.

THE CONTENT OF GENDER PLANNING TRAINING MODULES

Table A.1 identifies the four modules that form the core of the block-building process. Each module contains a number of components with materials adapted to meet the needs of different organizations. In addition, each stage introduces different training techniques, to diffuse confrontation and enable participants themselves to introduce gender dynamics. In outlining the components of each module, it is more important to identify the interrelationship between different elements, than to describe them separately. Therefore, the following description of each module addresses three questions:

1 What was done? – i.e., the content of the different components in each module;
2 How was it done? – i.e., the training techniques and materials used;
3 Why was it done? – i.e., the rationale underlying the different components.

Examples are drawn from several successful training courses undertaken in a variety of institutions to date.

Module I

Objective: To clarify fundamental concepts relating to gender, development and planning.

At the commencement of gender planning training, participants frequently arrive at the workshop with attitudes ranging from hostility and mistrust to irritation and scepticism. Either they do not believe in the importance of the issues, or are annoyed at being taken from their important work for such a 'trivial' issue. The workshop starts by asking participants to introduce themselves, identify their work responsibilities, and to mention their purpose and expectations in attending the workshop. Some participants have clear expectations, others are open-minded and have not really thought about it, while a third group may have come because their boss sent them. This introduction is an effective mechanism to diffuse initial tension. It also provides the trainers with important guidelines concerning objectives that they must ensure they address throughout the course of the workshop.

One trainer then introduces the workshop, mentioning briefly its purpose, the reason the organization considers such a workshop necessary, and the timetable and logistics. If the training occurs on the work site, it is particularly important to impress upon the participants that the tight schedule and structure demand full participation. Participants should not disappear into their offices during the course of the workshop.[3] The introduction usefully ends with a brief clarification of concepts such as sex and gender, sensitizing and training, and the separation of the political, personal and professional.

Gender planning training as an entry strategy chooses to assume that participants are gender-aware. Nevertheless, the first module is designed to ensure sensitization before developing skills for gender diagnosis. The technique starts with identification of the known. This allows participants to use their existing expertise or understanding to identify and discuss men's and women's roles in specific contexts in developing countries. Exercise One achieves this. As detailed in Table A.2 (see page 229), it requires participants to identify the various tasks undertaken over a twenty-four-hour period by the male and female household heads in a low-income family in different regions of the world. This work is undertaken in small groups. Tasks are plotted onto paper, with two axes representing the twenty-four-hour day of the household heads. A report-back follows, with charts put up for all to see. A member from each group takes everybody through their chart, describing very briefly the man's and woman's day. In the general discussion that follows, participants are invited to identify similarities and differences.

In the design of Exercise One, selection of households varies depending on the target group for training. For development agencies, as illustrated in Table A.2, a global range is appropriate. For an NGO or government agency working in one country, different sets of criteria can be introduced as appropriate to include further differences between rural and urban areas, income levels or household categories such as nuclear and extended. With training personnel from advanced industrial countries it is sometimes appropriate to introduce the example of a household from their country.

Throughout the workshop, small-group division for exercise work is undertaken in terms of pre-determined criteria, not in an arbitrary manner. These vary depending on the organization and relate to professional skill levels and areas of specialist expertise. Trainers therefore need brief particulars of the group before the workshop. In this exercise, the most important expertise is regional experience and the balancing of groups in terms of men and women. Training in the development community has highlighted what superficial knowledge most men, in practice, have of even the simplest divisions of labour within Third World low-income countries. Women, in contrast, tend to be much more aware of the issues, and others often encourage them to 'lead' this small-group work.

Exercise One provides the opportunity for participants themselves to identify important differences in the patterns of work between men and women. These relate to time, space, social interaction and type of activity, most of which can be identified globally, whatever the income level or location. In the discussion many issues emerge, of which the most important are the following. In terms of time, women work a longer day than men; they are the first up to prepare breakfast, and last to bed after completing other domestic tasks; women's days are fragmented as they change from one activity to the next, while men's days are characterized by blocks of time allocated to separate activities; men undertake single activities at a time, with women more likely to undertake a number of activities simultaneously, such as child care and agricultural work, cooking or water collection; where women go out to waged work, other female family members pick up their reproductive tasks; men's days divide between work periods and rest, while most women simply do not rest.

In spatial terms, in both urban and rural contexts, women usually only undertake productive work in the house or nearby, to balance it with reproductive work. Men travel further distances, especially given the differences in transport modes, with men more likely to use vehicular transport and women more likely to walk. Patterns of social interaction also differ. Men with free time socialize with other men; this can often include community level political activity. Women have less free time and more limited choice in social interaction. In many societies they socialize with other female family members, while undertaking other activities, such as shopping, collecting water, caring for children or community managing. Finally, the types of work differ: men who are more likely to be involved in visible work in the cash economy do not undertake household tasks. This is women's work, undertaken, alongside productive work, that includes subsistence and cash production. Where household-level production occurs a clear difference between men's and women's tasks exists.

This list could be endless, and although such an exercise may seem elementary, with the characteristics identified obvious, it is the first critical building block. It allows participants, rather than the trainer, to identify that because women and men play different roles within the family and community they undertake different activities. From here it is a simple step to recognize that women and men also have different needs. This exercise also diffuses 'cultural relativism' as cross-continent similarities far outweigh regional differences. As Table A.1 shows, the timing of the entire exercise is an hour. It is not intended to be a lengthy, drawn-out task; rather, it is an ice-breaker serving to sensitize participants.

Module II

Objective: To introduce the conceptual rationale of gender planning and the methodological tools to translate this into practice.

The objective of the second module it to provide skill transfer in gender diagnosis, and has two components. Leading on from Exercise One, and the discussion of roles and needs, is a formal introduction to the conceptual rationale for gender planning and the methodological tools to translate it into practice. Participants are introduced to the stereotypes in current planning in relation to family structure, models of the household and divisions of labour within it (see Chapter 2) and to gender planning tools such as gender-roles identification, gender needs assessment and the WID/GAD policy matrix (see Chapter 5). The training method can be either a formal lecture or a workshop presentation. This depends on the target group. In both cases it is accompanied by three handouts: on women's roles and practical and strategic needs (see Table 3.1, page 49); on different policies to Third World women (see Table 4.1, page 56); and definitions of concepts and tools used in gender diagnosis (see Table A.3, page 230).

The second component in this module is a short exercise, as detailed in Table A.4 (page 232). This provides participants with their first opportunity to apply the gender planning tools to selected projects. The purpose is to allow participants to identify and comparatively evaluate three different planning-intervention case studies, in terms of roles, needs and policy approaches. Several small groups undertake this work. Participants individually read the materials provided before discussing them, and then, after reaching agreement, they jointly complete the table provided (see Table A.5, page 233). The criteria for small-group selection at this stage are to balance expertise across groups, mixing together as skilfully as possible different status levels, professional disciplines and regional experience. A report-back session follows, in which the trainer requires each group to present its findings on each case study, followed by a general comparative discussion.

At this stage in the training, materials designed, or selected, specifically for this exercise are the most appropriate. The exercise requires three short case studies that can be quickly grasped, and provide comparative examples. Table A.6 (page 234) illustrates the guidelines for case studies. These suggest that the most useful are project-level examples that not only relate to the different roles of women, but also refer to different gender needs, reflect different policy approaches and cover as many sectors considered appropriate. Although NGOs commonly provide descriptions of their own projects, larger donors and government departments are less likely to have such small-scale project examples easily accessible. Therefore trainers may need to develop examples themselves. Table A.7 (page 235) provides examples of three short case studies that have proved particularly successful in training. This is both because of the important comparisons that can be made between them, as well as the content of each case study. It is useful to mention briefly a few issues that can be drawn out from the case studies, since similar training techniques are used again at a later stage in the training.

The three case studies describe income-generating projects. These differ not only in terms of their geographical location but also in terms of the roles on which they focus and the needs they meet. The puffed-rice project in Bangladesh is a classic example of an anti-poverty, income-generating project for women intended to focus on their productive role. In providing work for women in terms of existing skills, its intention is to meet very real PGNs for increased employment. In addition, by giving loans to women, the intention is also to meet the SGN of women to have greater control over their income. In reality, because of its design, the actual role on which it focuses is reproductive; it expects women to extend the time they already allocate to domestic cooking work. The result for women is more work, but no control over income, because men sell the product. Incidentally, they provide even more work for their wives by not separating rice and bran/husk. The project inadvertently results in increased control by men. Because it may improve household nutrition (if men hand over the profits they make), the project is more accurately identified as welfarist in approach.

In undertaking gender diagnosis, participants can identify the fundamental design problem as its problematic treatment of the household as the unit of production. This allows them to then identify the gender objective as ensuring that loans given to women remain in their control. Constraints identified may relate to men's objections and women's perceived inexperience, while the opportunities relate to the fact that women are more reliable in repaying loans. In devising a new strategy, one suggestion might be to ensure that bank accounts are in women's names, as well as holding meetings to explain to men how family income will increase.

An interesting contrast to this is another anti-poverty, income-generating project, the *gari* project in Ghana. Its focus is also on women's productive role and existing skills, but with change stimulated by the introduction of new technology. The project intends to meet the PGN for increased income, and achieves not only this but more. Because women control the technology, the production process and the marketing, their power relative to men increases (a potential SGN) while also creating male employment and income. Although the project was designed with an underlying anti-poverty approach, through the process of its implementation it ended up as an empowerment approach to WID. Again, gender objectives and alternative strategies can be suggested by participants.

The third project provides a further contrast because, as a 'classic' NGO project, it has manifold objectives, ranging from welfare through anti-poverty to empowerment. It focuses on all three roles of women: grandmothers provide child care in their reproductive role, women plant trees and create nursery gardens in their productive role, and ensure the success of the project in their community managing role. The intention is to meet the PGNs of fuel and food provision, as well as the SGN of changing divisions of labour by appointing a woman as project manager. Although the project was intended to involve both men and women, only women participated. The tool of roles identification is useful in understanding this. Because the project did not pay cash for production, men did not consider it 'men's work' and soon dropped out. Payment in food identified it as women's work, since in their reproductive role they have responsibility for family provisioning. Ultimately, men only participated as community politicians, in which they tried to prevent the introduction of a female project manager to the project. Ultimately, the dominant underlying rationale of the project was an efficiency approach, focusing on women as the most efficient planters of trees, and by that ensuring the project's success.

It is essential to stress that conclusions such as these are hypothetical in nature. When using these particular case studies in training, participants frequently reach very different conclusions, supported by highly valid arguments. In essence, case studies such as these are no more than training tools to enable participants to develop their capacity to undertake gender diagnosis. As this brief description shows, the exercise allows participants, while focusing on roles and needs, to address critical issues relating to the control of resources and the power of decision-making within the household and community. From this it is a very short step to identifying objectives, identifying constraints and opportunities, and then to designing new strategies.

The same exercise can be undertaken using a short film, rather than short case studies. Some gender planning training courses have used *The Lost Harvest*, a film about rice production in the Gambia. This portrays the impact of a new project designed to increase the agricultural production of both women and men in the area. Participants fill in a table while watching the film, and reach decisions in small groups before a report-back session. The film provides the opportunity to identify the roles of men and women, and divisions of labour before and after the project, women's and men's access to land, labour, savings, credit, skills and technology before and after the project, and finally any changes in gender needs met.

The advantages of films are the power with which visual images convey issues and that they allow participants to evaluate real projects. Nevertheless, the appropriateness and timing of the use of film in training has to be carefully assessed. The use of films requires time – the mere logistics of settling down to watch the video have to be taken into account – and, therefore, their use is more appropriate when training is undertaken at a slower pace than a one-day workshop allows. It also requires careful selection of appropriate film. The most successful are issue-based documentaries that highlight topics and concerns as opposed to training films with a message to teach audiences. Obviously, films can only convey a 'slice of life', however compelling. As a training tool, however, they lack the rigour, such as that provided by the three case studies cited above. However, they are a critical component in a longer training course. Not only do they change the pace of training but they also bring to life the issues that participants are addressing in their work. As such they can be built into different stages of longer courses to dramatic and powerful effect.

Module III

Objective: To apply gender planning tools to the organization's programmes and projects.

Once participants have grasped the principles of gender diagnosis and their use in short, case studies, it is time to turn to the reality of their own planning practice. This concerns both the work undertaken by their organizations generally and their specific work practices. The objective of both Modules III and IV, therefore, is to apply gender planning tools to the organization in which participants work. This is the most critical and difficult part of the training process. It divides into stages, the first of which (in Module III) is to apply the gender planning tools to the organization's policies, programmes or projects.

This is undertaken through Exercise Three, which in essence is similar to Exercise Two. The fundamental difference, as shown in Tables A.8 and A.9 (pages 239 and 240), is that it is based on documentation currently in use within the organization. Since such documents are often lengthy, participants receive them as preparation reading before the course. On a one-day course this can be up to a week in advance, while on a residential three-day course it provides evening work after the first training day. Exercise Three again comprises small group work – where dynamics are working well it is often a good idea to keep the same groups as allocated for Exercise Two. As a training technique it is important that all groups look at all materials. This ensures that in the report-back session each group can report its findings on each document in turn, followed by general discussion. In undertaking gender diagnosis participants conclude the exercise by noting the problems identified in the projects and programmes and by organizing and refining these problems into a cause-and-effect hierachy of problems.

As discussed in Chapter 8, the importance of using the organization's documentation as materials is that it places the training centrally to participants' work experience. This means that they cannot dismiss the exercise as irrelevant or inappropriate to the institution concerned. Quite frequently this happens with the previous exercise, when participants declare, 'We don't design projects like that!' This exercise is often the most revealing stage in the training workshop. Participants begin to recognize the gender-blindness, or confused WID approaches, in the documentation of their own organization.

Therefore the materials selected are critical, and collaboration with the organization's staff in the selection process is essential. Not only does this provide guidance as to what is available but also in decisions as to what is most suitable. Its success depends not only on the target group for training, but also on the objectives of the workshop and the time allocated to it. Different materials are selected for specific training groups to allow for both regional and sectoral variations, also the level at which groups are working. It is of little use training policy-level staff entirely with project-level documents, or the reverse. Finally, entire documents must be provided. Many organizations now have terms of reference that require project documents to include a section or paragraph referring specifically to women, while the rest of the document remains ungendered. Yet it is in the technical and financial sections of project documents that the most important gender diagnosis is necessary. If they are to use gender planning tools, participants need to be able to identify these.

One preparation task in gender planning training is the time-consuming task of reading as much of the organization's documentation as possible. Incidently, this also helps to familiarize trainers with the organization. Most commonly project- and programme-level documents are used. SIDA, for example, in its current three-day training course, has selected four documents as examples of four important stages in its planning process: appraisal, sector review, evaluation and special study. ODA, in its one-day course, uses three project appraisal documents which cover a variety of geographical and sectoral areas.

It is also important to ensure that document selection reflects the range of WID approaches. A gender-blind project document, with no reference at all to women and men, makes an important point. However, it is often more limited as a training tool than a confused gendered document. The latter provides participants with more positive training experience in the identification of gender roles and gender-needs assessment. Ultimately, a good set of materials is one that provides examples of both successful and unsuccessful documentation.

Module IV

Objective: To operationalize and institutionalize gender planning methodology; identification of the major constraints and opportunities.

The objective of Module IV is to continue the process of translating skills into planning practice by operationalizing and institutionalizing them into current work practices. In a one-day workshop there is only sufficient time to introduce the issues broadly, and to use examples provided by individual participants for a group-level discussion. As detailed in Table A.10 (page 241), Exercise Four (A) requires participants to decide what changes in operational procedures they might consider so as to ensure that their work incorporates a gender perspective.

In longer workshops, Module IV can be divided into different components that distinguish between operational and institutional issues, as separate components in the training process. Table A.11 (page 242) provides a prototype timetable for more detailed training on this module, based on a three-day workshop. It is useful to start the module with a formal presentation of gender planning procedures, as identified in Chapters 5 and 7. Following on from this is a workshop presentation by the institutional trainer outlining the way in which gender planning procedures can be incorporated into the planning cycle of the organization.

Participants then use this information on the relevance of different tools at various stages of the organization's planning cycle to undertake Exercise Four (B) (see Table A.12, page 243). Again in small groups, they are required to identify key prioritized 'gender objectives' on the basis of their gender diagnosis undertaken in the previous exercise. Following this is the identification of constraints and opportunities, and finally the development of an entry strategy which identifies the steps necessary to overcome the constraints and objectives (see Levy 1991 for further details on this). The chart provided (see Table A.13, page 244) allows participants to present the final version of their work in flip-chart form for presentation at the workshop. Cross-sectoral, regional and disciplinary expertise provides the basis for small-group selection.

In a three-day workshop it is often appropriate to change groups at this stage. This provides the opportunity to rectify any imbalances in ability, or intra-group tensions, that have appeared. Table A.11 (page 242) shows that, to be effectively undertaken, Exercise 4(B) takes a good half day to do.

In the second component of this module, the organization's trainers present issues in institutionalizing gender planning into the organization, such as those presented in Chapter 6. This allows participants to assess the relevance and importance of their WID office, and WID-assigned personnel in different parts of the organization, and to discuss the relative merits of mainstreaming and specializing institutional gender concerns.

The final component in the block-building is Exercise Five, outlined in Table A.14 (page 245). This provides the opportunity for participants to identify the strategies required to incorporate gender planning methodology into the redesign of their *current work*. This is similar to the previous exercise in many respects, except that it relates to current work practice, rather than to the institution's documentation, as in the previous exercise. It allows participants a second chance to undertake gender diagnosis, define gender objectives and develop entry strategies which include both the planning procedures and institutional structures that require changing if they are to be successful. As Table A.11 shows, in order to be effectively undertaken this exercise may take nearly a day. Even then, in the time available each group can only redesign the work of one of its members. The report-back needs to be put on a board or flip-chart so that it can be viewed comparatively by all at the end of the session.

At this stage it is necessary for the trainers to identify work problems of four participants, that individually and comparatively provide the greatest scope for small-group work. If participants all individually identify such 'problems' in written form at the end of the second day, trainers can select those that they would like to use in Exercise Five. As the last important exercise in the training process, the selection of small groups is critical. This is particularly important as most of the group help one member work on his or her 'problem'. As far as is possible, small groups should be based on common sectoral interests; the issues in an educational project interest participants if they have similar or associated concerns.

The final short exercise, as detailed in Table A.15 (page 246), brings together participants at similar levels or positions within the organization. They identify individually the procedures, both operational and institutional, they intend to adopt so as to integrate gender planning tools in their work. The concluding workshop provides the opportunity for each participant to discuss these procedures. Since recommendations made often require support from the WID office, this discussion provides the entry point for the confirmation of follow-up that is an essential part of the training process.

The most important issues to emerge through the different stages of Module IV are those of motivation and gender dynamics. Within any training group strong differences of opinion can emerge concerning the extent to which constraints are 'technical' or 'political' in nature. In a short workshop there is insufficient time to resolve conflicts, and the trainer's role must be to dissolve them as fast as possible. In a longer workshop there is more time to identify and elaborate the opportunities for change within the organization. Trainers then play an important role as catalyst in enabling consensus to be reached. A successful workshop is one in which participants consider they have gained 'technical' skills to improve their work practice. It is also one in which they place the underlying political issues on the organization's agenda. An evaluation sheet (see Table A.16, page 247) is provided at the end of the workshop. Participants return it to the trainers the following week. As Anderson (1991) has identified, this evaluation assesses the 'happiness quotient', with its main purpose being the redesign of the training workshop to meet the needs of particular target groups.

THE TRAINING OF TRAINERS

The final part of gender planning training is the training of trainers, which is an integral part of any training strategy. The description above highlights the trainer's role in the four modules of a gender planning training workshop. This shows the critical responsibility which trainers have in contributing to the relative success or failure of workshops. It also illustrates the level of skill sophistication required to run a difficult workshop with some twenty impartial or belligerent senior male and female professionals. Such training is at best hard work, at worst totally exhausting, with many skills best learned on the job.

The description of those who provide the training, in Chapter 8, identified two important facts: training is not teaching, and teachers are not necessarily trainers. In addition, it outlined the reasons for using senior staff from the organization as co-trainers. Although this is not a popular recommendation, comparative experience has proved its value as a long-term investment in institutionalizing and operationalizing gender planning.

The training process builds in the training of trainers, and comprises four clear stages. First, trainers undertake the training as regular workshop participants; secondly, they attend a training of trainers workshop especially designed to meet their needs; thirdly, they co-train alongside the consultants; fourthly, they take over the training process. As described in Chapter 8, some organizations choose to stop after stage three. Because of the combined power of 'inside' and 'outside' perspectives, they continue co-training with consultants. Others prefer to 'own' their own training and its associated methodology, adapting and changing to suit the needs of particular groups.

The training of trainers essentially involves the process of taking trainers through the issues outlined in this book, particularly those in Chapter 8 and this chapter. Consequently, it is only necessary to highlight very briefly the structure of a training of trainers workshop. As detailed in Table A.17 (page 248) it contains three modules. Its length depends both on the existing skills of the trainers and also on their knowledge of the gender planning methodology. With experienced trainers this can be undertaken in one day, although two days are preferable. Module I introduces trainers to the purpose of training and the components of a training strategy for gender planning. They then undertake an exercise where they develop an appropriate training strategy for an identified organization. Module II clarifies the purpose of the different modules, with many issues raised in this chapter discussed. Finally, in Module III, trainers themselves design a training course, identifying its components according to the different target groups defined by the trainers. While such a workshop provides trainers with the theoretical knowledge, ultimately it is the training experience that builds up their skills in the gender planning methodology.

In concluding this chapter on gender planning training it is important to stress that this is not in any way intended to provide a definitive picture of training. The description reflects my own personal voyage of discovery of gender planning methodology and training techniques, developed through a diversity of training workshop experiences. Since the first short course, 'Planning with Women for Urban Development', was run at the DPU London, and the first one-day workshop was undertaken with VSO in 1987, both the methodology and the training courses have undergone constant revision and adaptation as course participants have taken both the planning tools and training exercises back into their own environment. Because gender planning training is still in an exciting stage of development, the purpose of this chapter, ultimately, is simply to put on record the experience to date.

Table A.1   Timetable for prototype one-day gender planning training workshop

---

Module I   Objective: To clarify fundamental concepts relating to development, planning and gender

9.30	Introduction session: Development, planning, and gender
9.45	Exercise One: Daily activities of men and women in designated developing (and developed) countries
10.15	Report back and discussion
10.45	Coffee break

Module II   Objective: To introduce the conceptual rationale of gender planning and the tools in the gender planning methodology

11.00	Lecture or workshop: Critical issues in the theory and methodology of gender planning and their implications at the sector level
12.00	Exercise Two: To apply the gender planning tools to selected projects
12.45	Report back and discussion
13.15	Lunch break

Module III   Objective: To apply the tools in the gender planning methodology to selected work of the organization

14.15	Exercise Three: To apply the tools of the gender planning methodology to the organization's policies, programmes and/or projects
15.00	Report back and discussion
15.45	Tea

Module IV   Objective: To operationalize and institutionalize gender planning in participants' work context

| 16.00 | Operationalizing and institutionalizing gender planning in the current work of participants: final workshop |
| 17.00 | Close |

Source: C. Moser and C. Levy, Training materials developed for training in gender planning for development, 1984–90

Table A.2 Gender planning training workshop: Exercise One

Purpose: To identify the various daily tasks of men and women in low-income households in different regions of the world

1	Working in the groups allocated, discuss the daily lives of a husband and a wife in a low-income household in Africa, Asia or Latin America.
2	Decide on the location of your household (urban or rural) and specify the members of your household (including their age and sex).
3	Discuss the tasks that the husband and wife do on an average working day.
4	Using the paper provided, chart these tasks during a twenty-four-hour period.

Source: C. Moser and C. Levy, Training materials developed for training in gender planning for development, 1984–90

Table A.3 Definitions of concepts and tools used in gender diagnosis

---

Gender and sex

Sex identifies the biological difference between men and women. Gender identifies the social relations between men and women. It therefore refers not to men or women but to the relationship between them, and the way this is socially constructed. Gender relations are contextually specific and often change in response to altering economic circumstances.

Gender planning

A planning approach that recognizes that because women and men play different roles in society they often have different needs.

Gender roles

Gender planning recognizes that in most societies low-income women have a triple role: women undertake *reproductive, productive* and *community managing* activities, while men primarily undertake *productive* and *community politics* activities.

> *Reproductive role*: Child-bearing/rearing responsibilities, and domestic tasks done by women, required to guarantee the maintenance and reproduction of the labour force. It includes not only biological reproduction but also the care and maintenance of the workforce (male partner and working children) and the future workforce (infants and school-going children).

> *Productive role*: Work done by both women and men for pay in cash or kind. It includes both market production with an exchange-value, and subsistence/home production with actual use-value, and also potential exchange-value. For women in agricultural production this includes work as independent farmers, peasant wives and wage workers.

> *Community managing role*: Activities undertaken primarily by women at the community level, as an extension of their reproductive role, to ensure the provision and maintenance of scarce resources of collective consumption, such as water, health care and education. This is voluntary unpaid work, undertaken in 'free' time.

> *Community politics role*: Activities undertaken primarily by men at the community level, organizing at the formal political level, often within the framework of national politics. This is usually paid work, either directly or indirectly, through status or power.

Gender needs

Women have particular needs that differ from those of men, not only because of their triple role, but also because of their subordinate position in terms of men. It is useful to distinguish between two types:

> Practical gender needs (PGN) are the needs women identify in their socially accepted roles in society. PGNs do not challenge, although they arise out of, gender divisions of labour and women's subordinate position in society. PGNs are a response to immediate perceived necessity, identified within a specific context. They are practical in nature and often concern inadequacies in living conditions such as water provision, health care and employment.

> Strategic gender needs (SGN) are the needs women identify because of their subordinate position in society. They vary according to particular contexts,

related to gender divisions of labour, power and control, and may include such issues as legal rights, domestic violence, equal wages, and women's control over their bodies. Meeting SGNs assists women to achieve greater equality and change existing roles, thereby challenging women's subordinate position.

Policy approaches

Policy approaches to low-income Third World women have shifted over the past decade, mirroring shifts in macro-economic development policies. Five different policy approaches can be identified, each categorized in terms of the roles of women on which it focuses and the practical and strategic needs it meets.

Welfare: Earliest approach, 1950–70. Its purpose is to bring women into development as better mothers. Women are seen as passive beneficiaries of development. It recognizes the reproductive role of women and seeks to meet PGNs in that role through top-down handouts of food aid, measures against malnutrition and family planning. It is non-challenging and, therefore, still widely popular.

Equity: The original WID approach, used in the 1976–85 UN Women's Decade. Its purpose is to gain equity for women, who are seen as active participants in development. It recognizes the triple role, and seeks to meet SGNs through direct state intervention giving political and economic autonomy, and reducing inequality with men. It challenges women's subordinate position. It is criticized as Western feminism, is considered threatening and is unpopular with governments.

Anti-poverty: The second WID approach, a toned-down version of equity, adopted from the 1970s onwards. Its purpose is to ensure that poor women increase their productivity. Women's poverty is seen as a problem of underdevelopment, not of subordination. It recognizes the productive role of women, and seeks to meet the PGN to earn an income, particularly in small-scale, income-generating projects. It is most popular with NGOs.

Efficiency: The third, and now predominant, WID approach, adopted particularly since the 1980s debt crisis. Its purpose is to ensure that development is more efficient and effective through women's economic contribution, with participation often equated with equity. It seeks to meet PGNs while relying on all three roles and an elastic concept of women's time. Women are seen entirely in terms of their capacity to compensate for declining social services by extending their working day. Very popular approach.

Empowerment: The most recent approach, articulated by Third World women. Its purpose is to empower women through greater self-reliance. Women's subordination is expressed not only because of male oppression but also because of colonial and neo-colonial oppression. It recognizes the triple role, and seeks to meet SGNs indirectly through bottom-up mobilization of PGNs. It is potentially challenging, although its avoidance of Western feminism makes it unpopular except with Third World women's NGOs.

Source: C. Moser and C. Levy, Training materials developed for training in gender planning for development, 1984–90

Table A.4 Gender planning training workshop: Exercise Two

Purpose: To apply the gender planning tools to selected projects.

This exercise is not hypothetical but is based on brief case studies, illustrative of different interventions at the project level in developing countries.

| 1 | Working in the allocated groups, read the case studies provided and discuss them among yourselves. |
| 2 | Using the table provided, for each case study identify the following: |

(a) *Roles*:
On which of men and women's roles do you consider the intervention was intended to focus, and on which do you think it focused in practice?

(b) *Gender needs*:
Identify which gender needs the intervention was intended to meet, or met in practice.

(c) *Policy approaches to women*:
Identify the intended policy approach to women that underlies the agency's intervention. Was this reflected in practice?

Source: C. Moser and C. Levy, Training materials developed for training in gender planning for development, 1984–90

Table A.5 Chart to accompany Exercise Two: the application of gender planning tools to selected projects

Project	Role on which focused						Gender needs met				Policy approach to women
	Intention		Actual			Intention		Actual		
	R	CM/P	R	P	CM/P	PGN	SGN	PGN	SGN	
b										
a										
b										
a										
b										
a										
b										
a										

Source: C. Moser and C. Levy, Training materials developed for training in gender planning for development, 1984–90

Table A.6   Guidelines for short descriptions of project case studies for use in
            gender diagnosis

To undertake Exercise Two it is essential to have case studies of programmes or
projects concerned with low-income communities either appropriate to the organization
or supported by it. For authenticity these must be based as much as possible on actual
project experience. It is also helpful if these can relate to the experience of
training-course participants in terms of the sectors from which they are chosen. Case
studies must contain only description, since the purpose of the exercise is for
participants to provide their own analysis. In the selection of case studies to be used in
this exercise they should:

1   relate to the different roles of women;
2   refer to different gender needs;
3   reflect different policy approaches; and
4   cover as many sectors considered appropriate, identifying whether the
     institution undertaking the intervention is government, donor or NGO.

The following stereotypical projects may provide useful guidelines:

(a) Projects handing out food/services to women/children
1   relate to reproductive role;
2   meet practical gender needs for better child nutrition or health;
3   reflect a welfare approach to WID;
4   cover health and social welfare sectors.

(b) Projects involving women in traditional roles (such as sewing clubs)
1   relate to productive role;
2   meet practical gender needs for income and employment;
3   reflect anti-poverty approach to WID;
4   cover employment and economic development.

(c) Projects such as employment for women in 'men's jobs'
1   relate to productive roles;
2   meet practical gender needs for employment and strategic gender needs to
     change gender division of labour;
3   reflect an empowerment approach to WID;
4   cover employment and economic development.

(d) Projects that grant equality to women in legal land titles
1   relate to the triple role;
2   meet strategic gender needs for equal rights to land by challenging women's
     subordination;
3   reflect an equity approach to WID;
4   cover the legal sector.

Source: C. Moser and C. Levy, Training materials developed for training in gender planning
for development, 1984–90

Table A.7 Examples of case studies for use in Exercise Two

Case Study One

Puffed rice in Bangladesh:

In Bangladesh many agencies give small loans to individual women from landless families, to embark on a household-level rice-processing business. Equipment costs are low in such activities, with mainly household items used. However, the working capital needed to purchase unprocessed paddy is often beyond the means of very small families. Loans to overcome this problem can, therefore, help to provide employment for many women using existing skills and equipment.

One common rice-processing activity supported by such loans is the production of *muri* (puffed rice). Both men and women engage in this activity, the men in transportation and marketing and the women on the skilled production side. Men purchase paddy in the local market and take it to the women, who parboil it twice and dry it. The men then take it to the local rice mill, for milling. They return it to the farmsteads, where the women separate the grain from the mixed bran and husk. Although the mill delivers rice from one outlet, and bran/husk from another, the men normally load all of this into one bag. The women then painstakingly have to separate it out again, using a winnowing tray. The chaff with dried leaves is used for fuel in parboiling paddy and puffing rice, which demands great skill. The final product is then either sold locally or taken by bus to the major wholesale market in Dhaka. In either case it is the men who sell the final product and control the earnings.

Source: Marilyn Carr (1984) *Blacksmith, Baker, Roofing-sheet Maker... Employment for Rural Women in Developing Countries*, London: Intermediate Technology Publications, pp. 30–1.

Table A.7   (Continued)

Case Study Two

Gari processing in Ghana:

Gari (processed cassava) is becoming increasingly popular in Ghana because of the shortage of many other food items and because, once prepared, it is easy to cook. To prepare *gari*, however, is a very time-consuming, laborious business. It requires fermentation over several days, squeezing the water from the fermented cassava and finally roasting it over a wood fire.

To help women in a village in the Volta Region to increase their income through *gari*-processing, the Ghandian National Council on Women and Development introduced a new technology. The process involves a mechanical grater, a pressing machine to squeeze the water from the grated cassava and a large enamel pan for roasting. The pan holds ten times the volume of the traditional cassava pot. The system was developed locally, with advice on design from the women themselves. Before the introduction of this innovation, the women of the village produced fifty *gari* bags (weighing 50 kg each) every week. Now they are able to produce 5,000 to 6,000 bags a week. However, this increased output of *gari* can only be maintained with a higher yield of cassava in the area. Therefore, a male cassava growers' association formed to step up cassava production. The women's co-operative acquired a tractor to ensure that more land is put under cassava cultivation.

Involving the women in the design of the new technology undoubtedly contributed to the success of the project. It is interesting that the women obtained the funding to set up their male kin in a cassava production unit – perhaps men could learn from this example when the bottleneck is with processing capacity!

Source: Marilyn Carr (1984) *Blacksmith, Baker, Roofing-sheet Maker... Employment for Rural Women in Developing Countries*, London: Intermediate Technology Publications, pp. 18–19.

Table A.7 (Continued)

Case Study Three

Food for work nursery tree project:

This project was established in 1985. It was funded by a large NGO that also provided an adviser to ACCOMPLISH (Action Committee for the Promotion of Local Initiative in Self-Help), a local self-help group in Terekeka District of Equatoria Region, South Sudan.

The objectives of the project were as follows:

(i) to provide Terekeka District with a wide range of fruit, nuts, shade, forage and fuel-wood tree seedlings;

(ii) to provide a food income for Mundari people displaced from their home areas due to civil war;

(iii) to provide an ACCOMPLISH project to encourage Mundari women to play a more active role in the development of the Mundari community;

(iv) to create public awareness of desertification;

(v) to encourage soil and water conservation through tree planting;

(vi) to make tree planting an income-generating activity for women.

Three main nurseries were to be established – in Terekeka, Juba and Tali. Trees were then to be distributed throughout the district.

In November 1985 ACCOMPLISH appointed an agro-forestry adviser to advise on the design, establishment and daily running of the project, and to provide in-service training for a Mundari woman counterpart in ACCOMPLISH, and assistant staff. In January 1986 the Mundari woman counterpart was appointed. It was difficult to find a suitable person as few Mundari are able to acquire education. After her appointment, the counterpart underwent continuous in-service training and proved to be a great asset to the project. During the absence of the agro-forestry adviser, she took full responsibility for its running and proved to be very competent in managing every aspect of the nursery. The adviser recommended that she attend a Tree Nursery Management Course in Kenya. She was accepted and subsequently became a spokesperson for Sudanese women on agro-forestry and environmental issues.

Initially, members of the Food for Work Project were few in number, and work started in February 1986 with twelve women and ten men. However, members increased rapidly, 95 percent coming from the Muni area. Men worked on construction, building a watchman's *tukel*, a food/tool store and office, a nursery shade, fencing and drainage ditches. By March 1986, women made up 85 percent of the members. They prepared the land for planting, made and filled growing bags, planted trees and carried out the irrigation work. They also built a shade, to provide a place for the older women to sit and look after babies and young children while their mothers worked in the nursery. The number of displaced Mundari wanting to work continued to increase and the site had to be extended, with a further site established in June 1986.

An associated organization provided food for members, with a daily ration of 2 kg per person per day. By May 1986, a three-day rota had to be introduced to provide work for everyone, and the ration increased to 3 kg. By the end of May, they had planted 2,000 trees, but due to the worsening security situation, tree seedlings could not be

Table A.7   (Case Study Three continued)

transplanted throughout the district. They planted a further 800 fruit trees in July and since then many more. They also established a successful vegetable garden.

The Food for Work Tree Nursery Project has now been running successfully for eighteen months. The project was never intended to be a 'women's project'. However, by virtue of the nature of the work involved at the nursery, women have shown much more interest, willingness and understanding, and their participation has made up about 99 percent of the workers. The NGO adviser trained women, including a very competent project manager, to continue running the nursery after her departure in March 1987. However, the ACCOMPLISH committee initially raised objections to a female project manager and proposed a man for the post. Throughout the project members of the committee (all male) have shown little interest in the nursery, and have never visited the project area beyond the Project Office. The NGO convinced the Mundari that a female project manager should be appointed. It has to be seen whether the committee has a long-term interest in taking over the management of the nursery. The new project manager is coping well with the nursery, but says she would appreciate more support and understanding from ACCOMPLISH.

Source: C. Moser and C. Levy, Training materials developed for training in gender planning for development, 1984–90

Table A.8 Gender planning training workshop: Exercise Three

Purpose: To apply gender planning tools to the organization's programmes and projects.

1   This exercise is based on documentation given to participants for reading before the course. It covers three of the organization's projects.

2   Working in the allocated groups, and using the table provided, in each document identify the following:

(a)   *The potential impact of the project at the household level on*
       (i)   men and women in their productive role;
       (ii)   women in their reproductive role.

(b)   *The potential impact of the project at the community level on*
       (i)   women in their community managing role;
       (ii) men in their community politics role.

(c)   *Practical and strategic gender needs*
       i) Identify which gender needs each project intends to meet.

3   Note the problems that you identify in the project or programme. Complete your gender diagnosis by organizing and refining these problems into a cause-and-effect hierarchy of problems.

Source: C. Moser and C. Levy, Training materials developed for training in gender planning for development, 1984–90
C. Levy, Training materials for gender planning, 1990–92

Table A.9 Chart to accompany Exercise Three: the application of gender planning tools to the organizations, programmes and projects

Project	Potential impact at household level				Potential impact at community level		Potential gender needs met	
	Productive		Reproductive		Managing	Politics	PGN	SGN
	Men	Women	Women		Women	Men		

Source: C. Moser and C. Levy, Training materials developed for training in gender planning for development, 1984–90

Table A.10 Gender planning training: Exercise Four (A)

Purpose: To identify ways of operationalizing the gender planning methodology in your current work in the different departments of the organization

1	Select a task on which you are currently working.
2	Identify how far a gender planning methodology has been incorporated.
3	Based on the issues covered in the workshop, discuss how you might redesign your current work to incorporate gender planning procedures.
4	What are the major constraints and opportunities you expect to encounter?

Source: C. Moser and C. Levy, Training materials developed for training in gender planning for development, 1984–90

Table A.11 Gender planning training workshop: prototype timetable for more detailed training on Module IV

Day 1 (outlined in Table A.1)

Module IV Objective: To operationalize and institutionalize gender planning methodology: identification of the major constraints and opportunities

Day 2

10.45	A further introduction to gender planning procedures: gender objectives, gender consulation and entry points
11.30	Operationalizing gender planning procedures within the organization's planning cycle
12.00	Exercise Four (B): Redesigning of the organization's programmes and projects to incorporate gender planning procedures
12.30	Lunch
14.30	Exercise Four (B): Group work continued
15.15	Report back
16.00	Tea
16.15	Report back and discussion
17.30	Close

Day 3

8.30	Institutionalizing gender planning: resources and tools within both the WID Office and the organization
9.30	Discussion
9.45	Exercise Five: Strategies to operationalize and institutionalize gender planning procedures in participant's current work in the organization
10.15	Tea
10.30	Exercise Five: Group work continued
11.30	Report back
12.30	Lunch
13.30	The future of the gender planning methodology in the organization: where to next? Small-group discussion
14.00	Conclusion of the workshop: evaluation

Source: C. Moser and C. Levy, Training materials developed for training in gender planning for development, 1984–90
C. Levy, Training materials for gender planning, 1990–92

Table A.12 Gender planning training: Exercise Four (B)

Purpose: To operationalize gender planning procedures in the organization's policies, programmes and projects

1   The documentation you have already used illustrates different stages in the programme/project cycle. These are as follows:

    (a) appraisal;
    (b) socio-economic study;
    (c) annual sector review;
    (d) evaluation.

2   Working in the allocated groups, and using the project documents and the chart provided, undertake the following:

    (a) based on the hierarchy of problems in your gender diagnosis, identify key prioritized 'gender objectives' that will start the process of making the intervention more gender-aware (*what* to do);

    (b) identify the major constraints and opportunities you expect these 'working objectives' to encounter both in your organization and in the recipient country;

    (c) develop an 'entry strategy' to achieve the 'working objectives', identify the steps necessary to overcome/utilize the constraints and assets (*how* to do it).

Source: C. Moser and C. Levy, Training materials developed for training in gender planning for development, 1984–90
C. Levy, Training materials for gender planning, 1990–92

Table A.13   Chart to accompany Exercise Four (B): the redesigning of the organization's programmes or projects incorporating gender planning procedures

Project components of programmes	General objectives	Constraints	Opportunities	Entry strategy

Source: C. Levy, Training materials for gender planning, 1990–92

Table A.14 Gender planning training: Exercise Five

Purpose: To operationalize and institutionalize the gender planning procedures in the future work of the organization.

1   Working in the groups allocated, select a policy programme or project on which one member is currently working. On the basis of the issues covered in this workshop, initiate a process to better incorporate gender, by undertaking the following:

        (a) based on the hierarchy of problems in your gender diagnosis, identify key prioritized 'gender objectives' that will start the process of making the intervention more gender-aware (*what* do do);

        (b) identify the major constraints and opportunities you expect these 'working objectives' to encounter both in your organization and in the recipient country;

        (c) develop an 'entry strategy' to achieve the 'working objectives', identify the steps necessary to overcome/utilize the constraints and assets (*how* to do it).

Source: C. Moser and C. Levy, Training materials developed for training in gender planning for development, 1984–90
C. Levy, Training materials for gender planning, 1990–92

Table A.15   Gender planning training workshop: final small-group discussion

The future of gender planning methodology in the organization's work: where to next?

Purpose: To assess the application of the gender planning tools in your future work in the different departments and divisions of your organization.

Working in the groups allocated in terms of the different work responsibilities, specify how you will integrate the gender planning tools into your work.

Source: C. Moser and C. Levy, Training materials developed for training in gender planning for development, 1984–90

Table A.16 Gender planning training workshop: prototype evaluation of the
workshop

Please could you answer the following questions and return the completed evaluation
form as soon as possible to:

 The trainer
 The organization undertaking training

1   What were the positive aspects of the workshop?
      (In answering the question please provide details.)

2   What were the negative aspects of the workshop?
      (In answering the question please provide details.)

3   Which were the four most useful sessions (lectures/exercises)? Why?

4   What were the two least useful sessions (lectures/exercises)? Why?

5   Are there any issues from the workshop that you can integrate into your
      work? Please specify.

6   Was the workshop _____ too long?

                        _____ too short?

                        _____ the right length?

7   Any other comments/suggestions?

Name..

Department ....

Source: C. Moser and C. Levy, Training materials developed for training in gender planning
for development, 1984–90

Table A.17 Gender planning training: timetable of prototype training of trainers workshop

---

Module I Objective: To introduce the purpose of training and the components of a training strategy for gender planning

10.00	The purpose of training: your agenda; the gender planning training agenda
10.15	Background of gender planning training
10.30	Components of a training strategy: short exercise to develop an appropriate training strategy including target group; specification of the purpose of the training; the stages in the training process; length of individual workshops; number of trainers; budget
11.30	Coffee

Module II Objective: To clarify the purpose and use of the different modules in gender planning training workshops

11.45	The purpose and use of the different exercises; different formulations of the exercise; critical analytical issues; operational issues
12.45	The role of the lecture or workshop and other training tools
13.00	Lunch

Module III Objective: To develop training workshops for different target groups defined by the trainers

13.30	Exercise to design a training workshop for a target group in terms of timetable and workshop format; structure and content of sessions; materials required.
15.00	Report back
16.30	Tea and final discussion

Source: C. Moser and C. Levy, Training materials developed for training in gender planning for development, 1984–90

Notes

1 INTRODUCTION

1 Oakley, in her seminal writing on this issue in 1972, stated,

> 'Sex' is a biological term: 'gender' a psychological and cultural one. Common sense suggests that they are merely two ways of looking at the same division and that someone who belongs to, say, the female sex will automatically belong to the corresponding (feminine) gender. In reality this is not so. To be a man or a woman, a boy or a girl, is as much a function of dress, gesture, occupation, social network and personality as it is of possessing a particular set of genitals.
>
> (Oakley 1972: 158)

2 In a somewhat similar but more detailed definition Conyers (1982: 3) describes planning,

> not as an isolated activity but as part of a complex process of 'development' which involves a number of related activities, including:
>
> i) the identification of general goals or objectives;
> ii) the formulation of broad development strategies to achieve these objectives;
> iii) the translation of the strategies into specific programmes and projects;
> iv) the implementation of these programmes and projects; and
> v) the monitoring of their implementation and their impact in achieving the stated goals and objectives.

2 GENDER ROLES, THE FAMILY AND THE HOUSEHOLD

1 Machado (1987) provides a classic example of this problem in the case of a Brazilian housing project, intended for the poorest of the poor. In a context where some 30% of households were headed by women, the project identified potential beneficiaries as the *father* of the children. In reinforcing the planning stereotype of the nuclear family as the predominant household structure, these planners, even if inadvertently, excluded many of the poorest households headed by women from access to the project.

2 Rosenhouse, for instance, has defined the head as the person who provides the most 'effort on behalf of and commitment to the household' (1989: 12). Nevertheless, effort is defined in terms of market-orientated hours worked.

3 Longitudinal evidence from Guayaquil, Ecuador, shows that between 1978 and 1988 the number of extended households increased, as did the number of income earners within households, both necessary to enable households to cope with the increased costs of living during the period of recession and adjustment (see Moser 1992a).

4 A class review of Third World women-headed households is by Buvinic and Youssef, with Von Elm (1978). For more recent documentation, see Dwyer and Bruce (1988). In many contexts it is difficult to assess numbers accurately. Cultural discrimination and stigma against women-headed households make officials reluctant to admit the scale of the 'problem'. For instance, the author was informed by a senior Indian Administrative Service official that in a Metropolitan Indian city the rate was between 3 and 4%. A woman social worker working in the slums in that same city put the figure at 70%, demonstrating the differences in both definition and perception.

5 In trying to distil the complexities of this debate I have relied principally on the excellent analysis of the New Household Economics undertaken by Evans (1989), as well as recent work by Sen (1990), Hart (1990) and Young (1990).

6 The question posed by earlier feminists, once again reiterated, is, 'Why are both the neo-classical and Marxian paradigms so silent on the issues of inequality within the home' (Folbre 1986a: 251).

7 The term 'triple' has already been used in relation to women in various contexts. Bronstein (1982) discusses the three ways in which Third World peasant women suffer in terms of a 'triple struggle', 'as citizens of underdeveloped countries: as peasants, living in the most impoverished and disadvantaged areas of those countries; and as women in male-dominated societies'. In contrast, European feminists have used the term to refer to the increasing parental care-taking role of women (Finch and Groves 1983; Pascall 1986).

8 The gender division of labour and its ideological reinforcement are discussed historically by Scott and Tilly (1982). Barrett (1980) and Phillips (1983) provide a useful contemporary account for advanced economies. Amongst the succinct introductions to issues relating to women's subordination in an international perspective are those by Rogers (1980), Maguire (1984) and numerous articles in Young *et al.* (1981).

9 Because of the strong connotation that the term 'reproduction' still has to 'biological' reproduction and its links to sexuality and the reproduction of human life, there are many who still prefer to avoid the term. An interesting example is provided by the Commonwealth Expert Group on Women and Structural Adjustment who recently defined women as having four roles. Along with 'producers' and 'community organizers' they defined women's roles as 'home managers' and 'mothers', thus avoiding the term 'reproduction' and splitting it into two. However, the introduction of a gender ascriptive role, 'mother', alongside those relating to work or activities serves to confuse rather than clarify (Commonwealth Secretariat 1989).

10 Both Rogers (1980) and Oakley (1972) provide references to the diversity of anthropological evidence which indicates a continuum from societies with extremely rigid gender roles to those in which the distinction between masculine and feminine roles is highly flexible. Margaret Mead (1950), for example, in an attack on sexual determinism, identified examples of societies which insist that men be primarily responsible for infant and child care. She argued that while

Western societies see men as largely superfluous in pregnancy and childbirth, other societies believe them to be vitally concerned.

11 Studies of Third World women's employment include the edited volumes by Young and Moser (1981), Afshar (1985), Jain and Bannerjee (1985), Nash and Safa (1986), Redclift and Mingione (1985) and Menefee Singh and Kelles-Viitanen (1987).

12 I am grateful to Anne Phillips for this comment. It never ceases to strike me as the most succinct summary of the gender division of labour that I have yet heard.

13 In Western economies, the idea that women only worked to earn pocket money is clearly linked to capitalism and the struggle for the 'family wage' – the living wage needed to maintain a man, his wife and two children – with its consequent implications for the low wages of women as dependent wives (see Humphries 1977; Barrett and McIntosh 1980). Exported to the Third World, the same wage difference between male and female workers is also articulated in terms of men's need to earn the family wage. In describing the difference between male and female workers in clothing factories in Morocco, Joekes aptly states, 'Male workers and factory managers explain the difference in male and female earnings in terms of the phrase, "working for lipstick"' (1985: 183).

14 The term 'managing' rather than 'organizing' has been specifically selected to identify the activities of low-income women at the community level. This term reflects more accurately the manner in which women are involved informally in issues of collective consumption, than does the term 'organizing'. With its connotations of more formal organizations with their concomitant structures of power and control, organizing is largely outside the experience of women's community work.

15 In the 1970s and 1980s a major focus of concern was whether the popular struggles that arose from state intervention provide the basis for the development of urban social movements, in which broad class alliances are linked together in political struggle, not at the point of production, as was traditionally the case but at the point of residence. The fact that residential-level struggles are seen as inherently weaker than those around production issues is said to relate to the fact that issues of collective consumption do not necessarily coincide with class interests or antagonisms (Castells 1977, 1983). In this literature the role of women has mainly been mentioned descriptively in passing. If analysed conceptually it has been in terms of feminist consciousness, on the implicit assumption that through consumption-based struggles low-income women should develop an awareness of the nature of their gender subordination. I would argue that this contradicts with the reality in which women are reinforcing the gender divisions of labour through their role as community managers.

16 A number of impressive examples exist of women organizing successfully in gender-ascriptive roles. It was in their roles as the wives of imprisoned miner-union leaders that the comite de Amas de Casa de Siglo XX in Bolivia was formed after women had failed individually to get their husbands released. Some sixty women joined together, and travelled to the capital La Paz where they declared a hunger strike and published a manifesto. After the success of this spontaneous protest they formed the Housewives Committee within the labour union where they organized for issues of collective consumption such as improvements to housing and better education and medical services (Barrios de Chungara 1978). Of those women who have achieved success in the political domain, a considerable number, such as Eva Perón, Indira Gandhi and Corry Aquino have done so

in their gender-ascriptive role as daughters, wives or widows. For an interesting analysis of the so-called Supermadre phenomenon in the Latin American context, see Chaney (1979).

3 PRACTICAL AND STRATEGIC GENDER NEEDS AND THE ROLE OF THE STATE

1 Molyneux (1985a) does not define 'interests' as such; nor does she make this distinction between 'interests' and 'needs'.

2 Pascall (1986), Walby (1990), Showstack Sassoon (1987) and Kofman and Peake (1990) are all informative on women and the state in advanced economies, while Afshar (1987), Molyneux (1981, 1985b) and Moore (1988) provide useful introductions to the issue in the Third World.

3 Molyneux (1981) provides a useful working definition of socialism as referring to countries characterized by an expressed commitment to constructing a socialist country, and espousal of 'scientific socialism' (the principles of Marxism and Leninism), a high level of redistribution and the adoption of policies to abolish the private ownership of the means of production (p. 2).

4 The importance of this issue has been examined in such revolutionary contexts as Mozambique (Urdang 1989), Nicaragua (Tijerino 1978; Molyneux 1985a), South Africa (Kimbel and Unterhalter 1982; Beall *et al.* 1989) and El Salvador (Thomson 1986) as well as in the fundamentalist revolution in Iran (Afshar 1982; Nashat 1983).

5 See Kumari (1989) for a detailed discussion of dowry burning in India.

6 Nandita Haksar, in an excellent user-friendly manual for Indian women entitled *Demystification of Law for Women*, distinguishes between eight religious laws as follows: Hindu, Muslim, Christian, Jewish, Parsi, Ezhava, Naga customary and Tribal. As just one example of the differences between civil and religious law, she cites the fact that Hindu law is based on the older law of Manu which maintains that 'in childhood a female must be subject to her father, in youth to her husband... when her Lord is dead, to her sons' (Haksar 1986: 26). For an equally useful manual for Botswanan women which untangles the complexities for women of the differences between customary Tswana and common law, see Molokonne (n.d.).

7 The classic account of the impact of colonialism on women is by Etienne and Leacock (ed.) (1980). MacEwen Scott (1986) provides an interesting account of the manner in which stereotype assumptions about the role of women in LDCs have been influenced by the imposition of colonial administrative policy, the teaching of missions and more recently multinational labour legislation. For a comment on the debate relating to women's relative dependency on men and the state, see Moser (1989b).

8 For an extensive analysis of the consequences of household stereotypes for human settlement and housing policy, see Moser (1987a), while for detailed case-study examples, see Moser and Peake (1987).

9 Personal communication with official from the Ministry of Local Government, Botswana.

4 THIRD WORLD POLICY APPROACHES TO WOMEN IN DEVELOPMENT

1 Gender-blindness on the part of those formulating refugee programmes has severe consequences. As Harrell-Bond has commented, this has 'led to whole programmes going awry' (1986: 267).

2 In their critique of food aid, Jackson and Eade (1982) argue that MCH programmes can have detrimental effects on participants. They cite a survey in the Dominican Republic that found that food aid encouraged malnutrition. Pre-school children who ate rations at an MCH centre and were weighed monthly over two years were found not to gain weight noticeably except during the mango and avocado season and whenever food aid stopped. After questioning the mothers, the nutritionalist concluded that when children received food aid mothers tended to overestimate the value of this foreign 'wonder food' and to feed them less local food. Whenever the food aid failed to arrive, mothers would, as a matter of course, ensure that their children had food. This resulted in a weight gain. The experiment repeated elsewhere confirmed the same findings that with food aid there was no weight gain; without food aid weight increased.

3 UNICEF in a policy review paper on their agency's response to women's concerns showed a lack of clarity in defining their policy approach when they stated:

> There is a growing recognition within UNICEF of the multi-dimensional nature of women's roles and the need for increasing support to programmes that are both women- and mother-centred and are based on a development rather than a welfare approach.
>
> (UNICEF 1985: 4)

Although UNICEF distinguished here between 'welfare' and 'development', no further elaboration as to the definition of the latter was provided.

4 The *Oxford English Dictionary* defines 'equity' as 'fairness' and 'equality' as 'condition of being equal' (Fowler and Fowler 1964). Despite her important analysis of policy approaches to women, Buvinic (1983, 1986) has never sought to qualify the semantic shift from 'equity' to 'equality'. In much of the literature the two terms are often used interchangeably, despite the definitional differences.

5 It is important not to exaggerate the extent to which basic needs policy focused on women. ILO publications were very meagre on women, land reform was between (not within) households and women's burden was to be eased by water supplies, household technology, etc., focusing too on women's reproductive role (Ingrid Palmer, personal communication).

6 A review of a range of projects receiving financial assistance during 1987–88 from UK-based NGOs, Oxfam and Christian Aid, showed support for women's income-generating projects. At that time this included projects as diverse as handmills in Tanzania, fish-processing in Sierra Leone, dressmaking in Brazil, foundry work in Bangladesh, rope-making in Bangalore, India, khadi-spinning in Tamil Nadu, India, and clay pot-making in Indonesia. All were rural in focus. One largely unrecognized income-generating activity for low-income urban women is the street foods trade. Recent research has shown this is vital to urban survival strategies, with the widespread availability of low-cost cooked food reducing the time women spend in food-marketing and home preparation (Cohen 1986; UNICEF 1987).

7   The following quotations from international organizations illustrate not only that women are essential to the total development effort, but also suggest that efficiency has always been a primary rationale for working with women.

> The experience of the past ten years tells us that the key issue underlying the women in development concept is ultimately an economic one.
>
> (USAID 1982: 3)

> ... leaving questions of justice and fairness aside, women's disproportionate lack of education with its consequences in low productivity, as well as for the nutrition and health of their families, has adverse effects on the economy at large.
>
> (World Bank 1979: 2)

> Substantial gains will only be achieved with the contribution of both sexes, for women play a vital role in contributing to the development of their countries. If women do not share fully in the development process, the broad objectives of development will not be attained.
>
> (OECD 1983)

8   The project Development Alternatives with Women for a New Era (DAWN) grew 'from small seeds planted in Bangalore, India in August 1984' (DAWN 1985: 9). In 1985 collaborating institutions included the African Association of Women for Research and Development (AAWORD), Dakar, Senegal; the Women and Development Unit of the University of the West Indies (WAND), Barbados; the Asian and Pacific Development Centre (APDC), Malaysia; and the Chr. Michelson Institute (CMI), Norway, with support provided by the Ford Foundation, the Population Council and the Norwegian Agency for International Development (NORAD). Research organizations and funding institutions such as these have all played a critical role in supporting the development of the empowerment approach.

9   DAWN placed great importance on the relationship between international structures of domination of warfare and technology and the subordinate position of Third World women. Thus they state:

> We want a world where the massive resources now used in the production of the means of destruction will be diverted to areas where they will help to relieve oppression both inside and outside the home. This technological revolution will eliminate disease and hunger, and will give women means for safe control of her fertility.
>
> (1985: 73–5)

10 Considerable support has come from the governments of such countries as Canada, Denmark, Netherlands, Norway and Sweden. The Netherlands government, in their Development Cooperation Policy, has probably gone furthest in questioning the WID approach, and identifying the importance of what is termed an 'autonomy' approach (Boesveld *et al*. 1986). It is recognized that a stimulus for the shift in approach was the appointment of a feminist as Minister for Development Cooperation from 1982 until 1986 (Berden and Papma 1987).

5 TOWARDS GENDER PLANNING: A NEW PLANNING TRADITION AND PLANNING METHODOLOGY

1 Throughout this chapter the discussion on planning methodology - an often confusing and highly contradictory literature - refers extensively to the concise, analytical reviews by Healey (see Healey *et al.* 1982; Healey 1989).

2 This section develops further Caren Levy's categorization of five key principles of gender planning methodology (Moser and Levy 1986). In expanding this here, a further distinction is made between principles and tools for gender planning practice.

3 No clear consensus exists as to what is meant by participation, as is illustrated by three United Nations definitions. At one end of the continuum the UN in 1955 identified it with community development, whereby the active participation of the community creates a state of economic and social progress. At the other end of the continuum UNRISD in 1979 identified the objective of participation as that of 'increasing control over resources and regulative institutions in given social situations, on the part of groups and movements hitherto excluded from such control' (UNRISD 1979: 8). More specifically, in the field of human settlements, UNCHS in 1984 defined community participation as 'the voluntary and democratic involvement of the urban poor in carrying out these project activities' (UNCHS 1984: 1). For a further discussion of these issues, see Moser (1989b).

4 The distinction between the four participation questions of why, when, whose and how was originally developed in Moser (1989b). Many of the issues discussed in this section are drawn from this review article.

5 Sen (1990: 126) has commented that if a typical Indian rural woman was asked about her personal 'welfare' she would probably answer the question in terms of her reading of the welfare of her family.

6 In his recent work on public-sector accountability, Paul (1991) argues that public accountability is strengthened when government control is reinforced by public willingness and ability to exert *voice* (pressure to perform) or find *exit* (alternative sources of supply). Both these tactics are identified as balancing the phenomenon of 'capture' (the tendency of those who manage and control the allocation of public services to seek rents, not to serve public interest). He argues that the use of exit and voice by the public will depend on their relative cost and on the expected returns to the public from their use in the context of specific public services (1991).

6 THE INSTITUTIONALIZATION OF GENDER PLANNING

1 Examples to illustrate issues relating to the institutionalization of gender come from my professional experience of gender training during the 1980s. This included two semi-structured questionnaires undertaken with a diversity of professionals in the different organizations cited. The analysis does not pertain to be global or complete in so far as it can only describe part of what is a long-term process. Nevertheless, the range of organizations consulted provides a representative sample of the different types of organizations concerned to institutionalize gender.

2 Gordan's excellent case study, commissioned by the Commonwealth Secretariat,

focused on Jamaica, Guyana, Barbados, Grenada, Dominica and Belize. Its title, 'Ladies in Limbo', was intended to reflect 'the way in which Bureaux Heads perceived themselves performing the given task within their existing framework' (Commonwealth Secretariat 1985).

3   Countries included in the survey are Bahamas, Belize, Grenada, Guyana, Jamaica, St Kitts, St Lucia, St Vincent and Grenadines and Trinidad and Tobago. This survey was designed by the author and carried out by Sukey Field and Margaret Legum while undertaking a Gender Planning Training Workshop in Barbados in May 1990. The collaboration of participants, and of the Women and Development Unit at the Commonwealth Secretariat, is acknowledged.

4   See Chapter 8 and the Appendix for a more detailed description of training techniques, both generally and in relation to SIDA's experience.

5   A fascinating example of some of the contradictions between theory and practice is described by Button (1984) in her description of the Women's Units established in the mid-1980s in London borough councils. Typically, Women's Units were responsible for their own administration and clerical work, and each worker had equal status. While the women involved in establishing the units were concerned to make them non-hierarchical in keeping with feminist ideology, problems arose when an unstructured cell was located in a highly structured body. With fewer personnel to carry out their work, and a higher percentage of their time taken up with clerical work, staffing levels in the Women's Unit did not take account of the additional work created by self-administration such that arguably, 'these working arrangements actually lead to a waste of resources and restrict the capacity for specialist work' (1984: 47).

6   Freeman characterizes alternative organizations as containing the following principles: 'delegation of authority to specific individuals; the requirement that delegates shall be responsible to those who appoint them; distribution of power to avoid monopoly; rotation of tasks among individuals; allocation of tasks along rational criteria; diffusion of information; equal access to resources' (1974: 12).

7 OPERATIONAL PROCEDURES FOR IMPLEMENTING GENDER POLICIES, PROGRAMMES AND PROJECTS

1   The three priorities of equity, development and peace include: (1) *Equality* – legislative changes to promote equality, development and peace; power-sharing with women; setting up of institutions/procedures to monitor WID; removing stereotypes of women via education, media, etc.; sharing out domestic responsibilities in the family; integrating women's contribution into mainstream development; ratification of the Convention on Elimination of Discrimination against Women; (2) *Development* – increasing women's participation in all development areas; full involvement in political process; main focus on employment, health and education, improving opportunities for women, such as increasing grass-roots organizations of women, control of their own health and setting up child-care provision; (3) *Peace* – strengthening women's participation in peace activities and in independence movements; dismantling the arms race; promoting peace education; legal action to prevent violence against women; national machinery to be set up to combat domestic violence.

2   This section draws extensively from the excellent Third Monitoring Report on

the implementation of the DAC revised guiding principles on WID (OECD 1990).

3 In Canada, for example, WID has become one of its six major priorities in development effort. The other five are poverty alleviation, structural adjustment, environmental protection, food security and energy. In the Netherlands Development Cooperation Programme WID is one of four 'spearheads', the other three being poverty, environment and research.

4 Not only is epistemological sensitivity still acute, but language translation can also produce its own problems. Swedish SIDA, for example, in incorporating the gender planning methodology into their planning procedures, categorizes the five WID approaches as welfare, equity, anti-poverty, efficiency and 'strategic local initiatives' since there is no word for 'empowerment' in the Swedish language. When this is re-translated back into English, the meaning has subtly changed (SIDA 1990a).

5 The Netherlands government policy on women and development identifies eight specific objectives, which identify a strategic set of needs:

   1 To improve women's access to and control over production factors, services and infrastructure facilities.
   2 To reduce women's workload.
   3 To improve the enforcement of laws which lay down equal rights for women.
   4 To increase the involvement of women in decision-making in domestic, local, national and international levels.
   5 To improve the organization of women at all levels.
   6 To encourage the exchange of information and communication between women and women's groups, and change the stereotypical image of women.
   7 To improve women's knowledge and self-awareness.
   8 To combat physical violence and sexual abuse.

   (1989: 2)

6 Checklists for integrating women into development projects are now widespread, although they vary widely in content, form and depth. Those of ODA (n.d.), Christian Aid (1989), World Food Programme (1987), NORAD (1985) and Oxfam (n.d.) have been of particular use in this section.

7 To date, few agencies have published detailed planning tools. The following discussion relies heavily on the work of USAID (n.d., 1988) and Canadian CIDA (1986a, 1986b), both of which use amended versions of the Harvard case-study approach. In addition, the more recent work in progress of SIDA (1989) and ODA (n.d.; Moser and Levy 1989) employing the gender planning methodology is used to illustrate issues.

8 Recent sector-level project-cycle documents include the Operational Issues of the World Bank (with only the forestry sector completed to date (see Molner and Schreiber 1989)), and USAID's Gender Manual Series on basic education and vocational training (USAID 1986a), small-scale enterprises (USAID 1986b), agriculture and natural resources utilization (USAID 1987a).

9 The order of these activities is not standardized. Many agencies start with project objectives before identifying project beneficiaries. Obviously, if the target group were identified first it would be more probable that the project's objectives will meet their needs, with the reverse not necessarily being the case. Interestingly

enough, the UNDP short-format document does not even require identification of a project target group.

10 As Turbitt bluntly asserts:

> once a project has been forwarded to headquarters there is little chance that it will be sent back for reformulation, if it is found to be unresponsive to FAO WID policy. The best that can be expected is the addition of a 'women's' component which frequently operates apart from the mainstream activities, is late in starting up, and relies on relatively inexperienced APOs to save the cost of an expert. The more usual response is to make editorial changes to project documents, inserting references to women to give an appearance of compliance with the policy.
>
> (Turbitt 1987: 16)

11 An ex-head of the SIDA WID Office identified one of the ways in which the resident WID officer would find herself unintentionally marginalized from the visiting team was when decisions were reached over a beer, relaxing at night in the hotel bar.

12 This very brief description, of what is still a process in the making, comes from two sources: first, instructive training sessions given by Brita Ostberg and Carolyn Hannan-Andersson during SIDA gender training 1987–89, and secondly, SIDA publications (see SIDA 1989; Hannan-Andersson 1991). It is critical to emphasize that this is such a rapidly evolving process that those described here may soon be superseded by others. For instance, in the past year the Gender Office has placed less emphasis on country-specific plans, and focused more on sector-specific plans, because of their greater capacity to get to the action level (personal communication with Carole Hannan-Andersson, July 1992). Even though much of what is described in this case study may soon be out of date, its purpose is to highlight some of the processes agencies are involved in as they themselves develop and take further the gender planning methodology.

8 TRAINING STRATEGIES FOR GENDER PLANNING: FROM SENSITIZING TO SKILLS AND TECHNIQUES

1 The International Conference on Gender Training and Development Planning: Learning from Experience, co-hosted by the Population Council and the Chr. Michelson Institute, Bergen, Norway (12–15 May 1991).

2 For a concise review of the manner in which different donor institutions have adopted and adapted the gender analysis training, see Poats and Russo (1989) and Baele (1990).

3 This review, undertaken by Kathleen Cloud (1991) to coincide with the Bergen meeting, was based on the results of a questionnaire completed by fifty-five trainers, twenty of whom had experience across a range of institutions, while thirty-five reviewed the training efforts of an entire institution.

4 Cloud's (1991) review identifies three specific objectives which are categorized as increasing gender sensitivity, increasing responsiveness to women clients and improving implementation skills. While the fourfold typology developed here to some extent coincides with Cloud's categorization, it provides a more detailed disaggregation.

5 Interestingly enough, among more liberal groups this can create anxiety that an

increased focus on gender at the political level may result in an undesirable reduction in the focus on class as the major form of oppression. The concern here is that gender politics must be assumed within class politics.

6 As a low-paid NGO, Christian Aid has a high turnover of young, committed staff. In evaluating the early success of their gender planning training they recognized that staff turnover was itself one of the most effective ways of dealing with staff who will not take on the issue. Once training was institutionalized, it became accepted procedure for all staff to integrate it into their work.

9 TOWARDS AN EMANCIPATION APPROACH: THE POLITICAL AGENDA OF WOMEN'S ORGANIZATIONS

1 In a more detailed categorization Fowler (1988) distinguishes between Membership Organizations (MO) controlled, set up by and intended to benefit the members themselves; Private Service Organizations (PSOs) whose purpose is to promote the development of MOs; and Donor Local Organizations (DLOs) which are Southern branches of Northern NGOs, either as donors or implementing their own development programmes. Another categorization of the membership/service dichotomy, particularly in the Latin American context, is that between community-based groups (CBGs) and grass-roots support organizations (GSOs) (Lehmann 1990).

2 Cernea (1988) identifies these as including societal conflict and tension; the need to respond more effectively to crisis situations or new demands when traditional structures break down or become unresponsive; ideological and value differences with the powers-that-be in development planning and implementation; the realization that neither government nor the private business sector has the will or capacity to deal with certain acute social problems, and finally the determination to help people at the grass-roots to get organized and become involved in ongoing governmental development programmes.

3 I am grateful to Peter Sollis for highlighting the close interrelationship between social, economic and political and human rights work of NGOs, particularly in contexts of civil strife (see Sollis 1992, 1993).

4 Stein (1990) has identified the importance of the opening of 'political space' as a determinant of NGO activity, and the manner in which the closing of such space can result in NGO participatory service delivery models crossing the line that divides committed social action from political militancy. In this respect, the inherent contradiction between real empowerment and what happens to people once empowered, especially under repressive regimes, also requires consideration (Stein 1990).

5 This section of the book, in particular, has benefited from stimulating discussions with students doing the Gender, Development and Social Planning Option in the LSE MSc Course on Social Policy and Planning in Developing Countries. I would particularly like to acknowledge the contribution of Lulu Gwagwa, Anna Robinson and Helen Doyle.

6 Examples cited here come from the testimony of women participating in panels on 'The Role of Women in Ethnic Conflicts' and 'Violence: the Politics of Choice' at the Association for Women in Development Fifth International Forum, November 1991. These included women on different sides of the divide from Sri Lanka, Cyprus, Ireland, Israel, South Africa and the West Bank.

APPENDIX: GENDER PLANNING TRAINING: ITS METHODOLOGY AND CONTENT

1 Throughout this book it has been important to acknowledge the influence and collaboration of colleagues in developing gender planning. Nowhere is this more the case than in the development of the gender planning training. From the outset this was a collaborative endeavour, undertaken with co-trainers. In writing this chapter I am particularly conscious of the debt of recognition I owe to Caren Levy, my colleague and co-trainer in gender planning training from 1984 to 1990. Many of the exercises cited were painfully tried, tested and taken through a process of constant adaptation before they finally emerged in the form mentioned here. The process of tightening up one-day workshops so as to include within them the absolute maximum possible owes much to the resoluteness of Rosalind Eyben, the Senior Social Adviser at ODA, while the development of detailed training components on institutionalizing and operationalizing gender planning were greatly assisted by the supportive articulation of training needs of both Brita Ostberg and Carolyn Hannan-Andersson at SIDA. Finally, Sukey Field helped me subtly to introduce more gender dynamics into gender planning training.

2 The purpose of providing a 'prototype' training workshop, is to show the way in which the four different modules can be timetabled into a one-day workshop. As a 'prototype' this combines elements of different workshops undertaken, as well as reflecting the analysis and research undertaken in writing this book.

3 This somewhat minor point cannot be over-emphasized. Because of the tightly structured participatory nature of this training methodology, nothing is more disruptive than participants disappearing in the middle of training for a 'more important' matter. Interestingly enough, in my experience, NGO participants are more guilty of this offence than the more disciplined participants from larger bureaucracies such as bilateral donors. Where possible, it is useful to identify this as a potential problem at the outset of training, and request those concerned to come for a future session when they can allocate the entire time required. On more than one occasion at which this technique has been employed participants have voluntarily returned at a later date for training.

Bibliography

Afshar, H. (1982) 'Khomenini's teachings and their implications for women', *Feminist Review*, 12: 59–62.
—— (ed.) (1985) *Women, Work and Ideology in the Third World*, London: Tavistock.
—— (ed.) (1987) *Women, State and Ideology*, London: Macmillan.
Afshar, H. and Dennis, C. (eds) (1992) *Women and Adjustment Policies in the Third World*, Basingstoke: Macmillan.
Agarwal, B. (1981) *Water Resources Development and Rural Women*, New Delhi: Ford Foundation.
—— (1986) *Cold Earth and Barren Slopes: Woodfuel Crises in the Third World*, California: Riverdale.
Aklilu, D. (1991) *Gender Training: Experiences, Lessons and Future Directions. A UNIFEM Review Paper*, New York: UNIFEM.
Andersen, C. and Baud, I. (eds) (1987) *Women in Development Cooperation: Europe's Unfinished Business*, Antwerp: Centre for Development Studies.
Anderson, M. (1990) *Women on the Agenda: UNIFEM's Experience in Mainstreaming with Women 1985–1990*, New York: UNIFEM.
—— (1991) 'Does gender training make a difference? An approach to evaluating the effectiveness of gender training', Mimeo.
Anderson, M. and Chen, M. (1988) 'Integrating women or restructuring development', *Association of Women in Development*, Occasional Paper prepared for WID Colloquium on Gender and Development, Washington, DC: AWID.
Antrobus, P. (1989) 'Women and planning: the needs for an alternative analysis', Paper presented at Women, Development Policy and the Management of Change seminar, Barbados.
—— (1991) 'Development alternatives with women', in *The Future for Women in Development; Voices from the South*, Proceedings of the Association of Women in Development colloquium, Ottawa: The North, South Institute.
Baele, S. (1990) *Gender and Development: Elements for a Staff Training Strategy*, Geneva: International Labour Organization.
Balayon, T. (1991) 'Gender dynamics: a conceptual framework', Paper presented at the International Conference on Gender Training and Development Planning: Learning from Experience, Bergen, Norway (12–15 May 1991).
Barrett, J., Dawber, A., Klugman, B., Obery, I., Shindler, J., and Yawitch, J. (1985) *South African Women on the Move*, London: Zed.
Barrett, M. (1980) *Women's Oppression Today*, London: Verso.

Barrett, M. and McIntosh, M. (1980) 'The family wage', some problems for socialists and feminists', *Capital and Class*, 11.

Barrig, M. and Fort, A. (1987) 'La ciudad de las mujeres: pobladoras y servicios y el caso de El Augustino', *Women, Low-Income Households and Urban Services Working Papers*, Lima.

Barrios de Chungara (1978) *Let Me Speak: Testimony of Domitila, a Woman of the Bolivian Mines*, New York: Monthly Review Press.

Baum, W.C. (1982) *The Project Cycle*, Washington, DC: The World Bank.

Beall, J., Hassim, S., and Todes, A. (1989) 'A bit on the side?' Gender struggles in the politics of transformation in South Africa', *Feminist Review*, 33.

Becker, G. (1965) 'A theory of the allocation of time', *Economic Journal*, 75.

Beneria, L. (1979) 'Reproduction, production and the sexual division of labour', *Cambridge Journal of Economics*, 3.

—— (ed.) (1982) *Women and Rural Development*, New York: Praeger.

Berden, M. and Papma, A. (1987) 'The Netherlands' in C. Anderson and I. Baud (eds), *Women in Development Cooperation: Europe's Unfinished Business*, Antwerp: Centre for Development Studies.

Bernheim, B.D. and Stark, O. (1988) 'Altruism within the household reconsidered: do nice guys finish last?', *American Economic Review*, 78: 1034–45.

Beveridge, W. (1942) *Social Insurance and Allied Services*, London: HMSO, Cmnd 6404.

Bhatty, Z. (1980) 'Economic roles and status of women: a case study of women in the Beedi industry in Allahabad', *ILO Working Paper*, Geneva: International Labour Organization.

Blau, P.M. (1956) *Bureaucracy in Modern Society*, New York: Random House.

Blumberg, R.L. (1988) 'Income under female vs male control: differential spending patterns and the consequences when women lose control of returns to labour', Draft report prepared for the World Bank, Washington, DC.

Boesveld, M., Helleman, C., Postel-Coster, E., and Schrijvers, J. (1986) 'Towards autonomy for women: research and action to support a development process', *Working Paper No. 1*, The Hague: RAWOO.

Bonepath, E. (1982) 'A framework for policy analysis', in E. Bonepath (ed.), *Women, Power and Policy*, New York: Pergamon.

Bonnerjea, L. (1985) *Shaming the World: The Needs of Refugee Women*, London: World University Service.

Boserup, E. (1970) *Woman's Role in Economic Development*, New York: St Martins Press.

Boxer, M. (1982) 'For and about women: the theory and practice of women's studies in the Unites States', in N. Keohane, M. Rosaldo and B. Gelpi (eds), *Feminist Theory: A Critique of Ideology*, pp. 237–71, Brighton: Harvester Press.

Bronstein, A. (1982) *The Triple Struggle*, London: War on Want.

Bruce, J. (1980) *Market Women's Co-operatives: Giving Women Credit*, New York: Population Council.

Bujra, J. (1986) 'Urging women to redouble their efforts...: Class, gender and capitalist transformation in Africa', in C. Robertson and I. Berger (eds), *Women and Class in Africa*, pp. 117–40, New York: Africana Publishing Company.

Bunch, C. (1980) 'Copenhagen and beyond: prospects for global feminism', *Quest*, 5.

Button, S. (1984) 'Women's committees: a study of gender and local government policy formation', *Working Paper 45*, Bristol: University of Bristol, School for Advanced Urban Studies.

Buvinic, M. (1982) 'Has development assistance worked? Observations on programs for women in the Third World', Paper presented at annual meeting of the Society for International Development, Baltimore, MD.

—— (1983) 'Women's issues in Third World poverty: a policy analysis', in M. Buvinic, M. Lycette, and W. McGreevey, *Women and Poverty in the Third World*, Baltimore: Johns Hopkins University Press.

—— (1986) 'Projects for women in the Third World: explaining their misbehaviour', *World Development*, 14(5).

Buvinic, M., Youssef, N. with Von Elm, B. (1978) 'Women-headed households: the ignored factor in development planning', Report submitted to the Office of Women in Development, Agency for International Development, Washington, DC: International Center for Research on Women.

Canadian International Development Agency (CIDA) (1986a) *Guidelines for Integrating WID into Project Design and Evaluation*, Ottawa: CIDA.

—— (1986b) *Women in Development and the Project Cycle: A Workbook*, Draft document prepared by O. Navia-Melbourn and J. MacKenzie, Ottawa: CIDA.

—— (1988) *A Vital Force in Development: Report on CIDA's Progress in Implementing its Women in Development Action Plan*, Ottawa: CIDA.

Caplan, P. (1978) 'Women's organizations in Madras City, India', in P. Caplan and J. Bujra (eds), *Women United, Women Divided*, London: Tavistock.

—— (1985) *Class and Gender in India: Women and their Organizations in a South Indian City*, London: Tavistock.

Cardon, M.L. (1974) 'Women's liberation: organization', in *The New Feminist Movement*, New York: Russell Sage Foundation.

Carr, M. (1984) *Blacksmith, Baker, Roofing – Sheet Maker... Employment for Rural Women in Developing Countries*, London: Intermediate Technology Publications.

Castells, M. (1977) *The Urban Question*, London: Edward Arnold.

—— (1983) *The City and the Grassroots*, London: Edward Arnold.

Cernea, M. (1988) 'Nongovernmental organizations and local development', *World Bank Discussion Paper No. 40*, Washington, DC: World Bank.

Chaney, E. (1979) *Supermadre*, Austin, Texas: University of Texas Press.

Chen, L.C., Huo, E. and D'Souza, S. (1981) 'Sex bias in the family allocation of food and healthcare in rural Bangladesh', *Population and Development Review*, 1(1).

Christian Aid (1987) *Women and Christian Aid: Paper Two Adding Strength and Vision*, London: Christian Aid.

—— (1989) *Gender Guidelines*, London: Christian Aid.

Cloud, K. (1991) 'Gender training: the state of the art, 1991', Paper presented at the International Conference on Gender Training and Development Planning: Learning from Experience, Bergen, Norway (12–15 May 1991).

Cohen, M. (1986) 'Women and the urban street food trade: some implications for policy', *DPU Gender and Planning Working Paper no. 12*, London: Development Planning Unit.

Commonwealth Secretariat (1985) *Record of the Workshop on Ladies in Limbo Revisited*, London: Commonwealth Secretariat Women and Development Programme.

—— (1988a) *Report to the Secretary-General by the Secretariat Committee on Women and Development*, London: Commonwealth Secretariat.

—— (1988b) *The Convention on the Elimination of All Forms of Discrimination*

Against Women: The Reporting Process – A Manual for Commonwealth Jurisdictions, London: Commonwealth Secretariat.

—— (1989) *Engendering Adjustment for the 1990s*, London: Commonwealth Secretariat.

Conference of Socialist Economists (CSE) (1976) 'On the political economy of women', *Pamphlet No. 2*, London: Conference of Socialist Economists.

Conyers, D. (1982) *An Introduction to Social Planning in the Third World*, Chichester, John Wiley and Sons.

Cornia, G., Jolly, R., and Stewart, F. (1987) *Adjustment with a Human Face*: Vol. 1, Oxford: Oxford University Press.

—— (1988) *Adjustment with a Human Face*: Vol. 2, Oxford: Oxford University Press.

Dankelman, I. and Davidson, J. (1988) *Women and Environment in the Third World: Alliance for the Future*, London: Earthscan Publications with IUCN.

Davin, D. (1987) 'Gender and population in the People's Republic of China', in H. Afshar (ed.), *Women, State and Ideology*, London: Macmillan.

Desai, M. (1990) 'Plural meanings of women's activism in India', Pittsburgh Women's Anthropology Group (eds), *International Women's Anthropology Conference Newsletter*, 12 (Winter).

Development Alternatives with Women for a New Era (DAWN) (1985) *Development, Crisis, and Alternative Visions: Third World Women's Perspectives*, Delhi: DAWN.

Development Alternatives (1987) Evaluation of the International Center for Research on Women Cooperative Agreement Program with AID PPC/WID, Washington: Development Alternatives.

Dey, J. (1981) 'Gambian women: unequal partners in rice development projects', in N. Nelson (ed.), *African Women in the Development Process*, London: Frank Cass.

Diamond, L. (1989) 'Beyond authoritarianism and totalitarianism: strategies for democratization', *Washington Quarterly* (Winter).

Dixon-Mueller, R. (1985) 'Women's work in Third World agriculture', *ILO Women, Work and Development*, no. 9, Geneva: International Labour Organization.

Dwyer, D. and Bruce, J. (1988) *A Home Divided: Women and Income in the Third World*, Stanford, CA: Stanford University Press.

Edholm, F., Harris, O., and Young, K. (1977) 'Conceptualizing women', *Critique of Anthropology*, 3(9/10).

Editorial Collective (1987) *In Search of Our Bodies: A Feminist Look at Women, Health and Reproduction in India*, Bombay: Shakti.

Elson, D. (1991) 'Male bias in macro-economics: the case of structural adjustment', in D. Elson (ed.), *Male Bias in the Development Process*, Manchester: Manchester University Press.

Elson, D. and Pearson, R. (1981) 'Nimble fingers make cheap workers': an analysis of women's employment in Third World export manufacturing', *Feminist Review*, 7 (Spring).

Enabulele, A. (1985) 'The role of women's associations in Nigeria today', Mimeo.

Erinle, M.T. (1986) 'Evaluation of a government income generating project for women in Kwara, Nigeria', in 'Planning with Women for Urban Development 1985 Participants' Reports', *Gender and Planning Working Paper no. 9*, London: Development Planning Unit.

Esman, M. and Uphoff, N. (1984) *Local Organizations: Intermediaries for Rural Development*, Ithaca, NY: Cornell University Press.

Etienne, M. and Leacock, E. (eds) (1980) *Women and Colonization: Anthropological Perspectives*, New York: Praeger.

Evans, A. (1989) 'Women, rural development and gender issues in rural household economics', *Discussion Paper 254*, Sussex: Institute of Development Studies.

Evans, A. and Young, K. (1988) *Gender Issues in Household Labour Allocation – the Case of Northern Province, Zambia*, Report to ESCOR, London: ODA.

Feldman, R. (1989) *Women for a Change: The Impact of Structural Adjustment on Women in Zambia, Tanzania and Mozambique*, London: War on Want Publications.

Feldstein, H. (1986) 'Intrahousehold dynamics and farming systems research and extension: conceptual framework', Mimeo.

Fernando, M. (1987) 'New skills for women: a community development project in Colombo, Sri Lanka', in C.O.N. Moser and L. Peake (eds), *Women, Human Settlements and Housing*, London: Tavistock.

Finch, J. and Groves, D. (eds) (1983) *A Labour of Love: Women, Work and Caring*, London: Routledge and Kegan Paul.

Folbre, N. (1986a) Hearts and spades: paradigms of household economics', *World Development*, 14(2), pp. 245–55.

—— (1986b) 'Cleaning house: new perspectives on households and economic development', *Journal of Development Economics*, (June).

Food and Agricultural Organization (FAO) (1987) 'Identification of regular programme beneficiaries by gender: an analysis of the Nov. 1986 LANSYS Date', Mimeo.

Fort, A. (1991) 'The Peruvian experience in gender training: description and analysis', Paper presented at the 'International Conference on Gender Training and Development Planning: Learning from Experience', Bergen, Norway (12–15 May 1991).

Fowler, A. (1988) 'Non-Governmental Organizations in Africa: achieving comparative advantage in relief and micro-development,' *Institute of Development Studies Discussion Paper no. 249*, Brighton: Institute of Development Studies.

Fowler, H. and Fowler, G. (eds) (1964) *The Concise Oxford Dictionary of Current English*, Oxford: Oxford University Press.

Freeman, J. (1974) 'The tyranny of structureless', in J.Jaquette (ed.), *Women in Politics*, New York: John Wiley.

Gardiner, J. (1977) 'Women in the labour process and class structure', in A. Hunt (ed.), *Class and Class Structure*, London: Lawrence and Wishart.

Germaine, A. (1977) 'Poor rural women: a policy perspective', *Journal of International Affairs*, 30.

Ghai, D. (1978) 'Basic needs and its critics', *Institute of Development Studies Bulletin*, 9(4).

Goddard, V. (1981) 'The leather trade in the Bassi of Naples', *Institute of Development Studies Bulletin*, 12(3).

Gomez, M. (1986) 'Development of women's organizations in the Philippines', in 'Women, struggles and strategies: Third World Perspectives', *Women's Journal*, no. 6, Rome: ISIS International.

Gordan, S. (ed.) (1984) *Ladies in Limbo: The Fate of Women's Bureaux*, London: Commonwealth Secretariat.

Government of Jamaica, Bureau of Women's Affairs (1987) *National Plan of Action*, Jamaica: Government of Jamaica.

Grindle, M. (ed.) (1980) *Politics and Policy Implementation in the Third World*, Princeton, NJ: Princeton University Press.

Gunn, L.A. (1978) 'Why is implementation so difficult?', *Management Services in Government* (November).

Guyer, J. and Peters, P. (1987) Introduction in 'Conceptualizing the household: issues of theory and policy in Africa', *Development and Change*, 18(2), pp. 197–214.

Haksar, N. (1986) *Demystification of Law for Women*, New Delhi: Lancer Press.

Hambleton, R. (1986) *Rethinking Policy Planning*, University of Bristol: School for Advanced Urban Studies.

Hannan-Andersson, C. (1991) 'Experiences with gender training – how did it work and how was it used?: some experience from the Swedish International Development Authority 1988–1991', Paper presented at the International Conference on Gender Training and Development Planning: Learning from Experience, Bergen, Norway (12–15 May 1991).

—— (1992) 'Gender planning methodology: three papers on incorporating the gender approach in development cooperation programmes', *Rapporter Och Notiser 109*, Lund: University of Lund.

Hardiman, M. and Midgley, J. (1982) *The Social Dimensions of Development*, London: Wiley.

Harrell-Bond, B.E. (1986) *Imposing Aid: Emergency Assistance to Refugees*, Oxford: Oxford University Press.

Harris, O. (1981) 'Households as natural units', in K. Young, C. Wolkowitz and R. McCullagh (eds), *Of Marriage and the Market*, London: CSE.

Hart, G. (1990) 'Imagined unities: constructions of "the household" in economic theory', Paper prepared for the Economic Anthropology Conference, University of Arizona, Tucson, 27–29 April.

Hartmann, H. (1981) 'The unhappy marriage of Marxism and feminism: towards a more progressive union', in L. Sargent (ed.), *Women and Revolution*, Boston: South End Press.

Healey, P. (1989) 'Planning for the 1990s', Mimeo.

Healey, P., McDougall, G., and Thomas, M. (1982) *Planning Theory: Prospects for the 1980s*, Oxford: Pergamon.

Helzner, J. and Shepard, B. (1990) 'The feminist agenda in population private voluntary organizations', in K. Staudt (ed.), *Women, International Development and Politics*, Philadelphia: Temple University Press.

Hilsum, L. (1983) 'Nutrition education and social change: a women's movement in the Dominican Republic', in D. Morley, J. Rhode and G. Williams (eds), *Practising Health for All*, Oxford: Oxford University Press.

Himmelstrand, K. (1990) 'Can an aid bureaucracy empower women?' in K. Staudt (ed.), *Women, International Development and Politics*, Philadelphia: Temple University Press.

Hirschman, A. (1984) *Getting Ahead Collectively: Grass Roots Experiences in Latin America*, Elmsford, NY: Pergamon Press.

Hirschmann, D. (1990) 'The Malawi case: enclave politics, core resistance and Nkhoswe No. 1', in K. Staudt (ed.), *Women, International Development and Politics*, Philadelphia: Temple University Press.

Holden, P. (1988) *Constraints on Increasing the Proportion of Women Benefiting from ODA Funded Training Awards: Executive Summary*, London: Overseas Development Administration.

Holmquist, F. (1984) 'Self-help: the state and peasant leverage in Kenya', *Africa*, 54(3): 72–92.

Howard-Borjas, P., Karl, M., and Spring, A. (1991) 'Gender analysis workshop for professional staff: FAO's mid-term review of lessons learned', *Working Paper Series no. 7*, Food and Agricultural Organization, Rome.

Humphries, J. (1977) 'Class struggle and the persistence of the working class family', *Cambridge Journal of Economics*, 1(3).

Huxley, M. (1988) 'Feminist urban theory: gender, class and the built environment', *Transition* (Winter), 39–43.

INSTRAW (1988) See United Nations Institute for Training and Research for the Advancement of Women.

International Women's Tribune Centre (IWTC) (1985) 'Women, money and credit', *Newsletter 15*, New York: IWTC.

Jackson, T. and Eade, D. (1982) *Against the Grain: The Dilemma of Project Food Aid*, Oxford: Oxfam.

Jagger, A. (1977) 'Political philosophies of women's liberation', in M. Vetterling-Braggin, F. Elliston and J. English (eds), *Feminism and Philisophy* (pp. 5–21), Totowa, N.J: Littlefield, Adams and Company.

Jain, D. and Banerjee, N. (eds) (1985) *Tyranny of the Household*, New Delhi: Shakti.

Jayawardena, K. (1986) *Feminism and Nationalism in the Third World*, London: Zed.

Joekes, S. (1985) 'Working for lipstick? male and female labour in the clothing industry in Morocco', in H. Afshar (ed.), *Women, Work and Ideology in the Third World*, London: Tavistock.

Jolly, R. (1987) 'Women's needs and adjustment policies in developing countries', Address given to Women's Development Group, OECD, Paris (mimeo).

Jumani, U. (1987) 'The future of home-based production', in A. Menefee Singh and A. Kelles-Viitanen, *Invisible Hands*, New Delhi: Sage.

Kaplan, T. (1982) 'Female consciousness and collective action: the case of Barcelona, 1910–1918', *Signs*, 7, 3.

Karl, M. (1983) 'Women and rural development', in *Women in Development: a Resource Guide*, Geneva: ISIS Collective.

Khairy, H. (1986) 'Community participation in urban development projects in Amman, Jordan', in 'Planning with women for urban development 1984, Participants' Reports', *Gender and Planning Working Paper No. 8*, London: Development Planning Unit.

Kimble, J. and Unterhalter, E. (1982) 'We opened the road for you, you must go forward: ANC women's struggles, 1912–1982', *Feminist Review*, 12: 11–35.

Kofman, E. and Peake, L. (1990) 'Into the 1990s: a gendered agenda for political geography', *Political Geography Quarterly*, 9(4): 313–36.

Kumar, R. (1989) 'Contemporary Indian feminism', *Feminist Review*, no. 33.

Kumar, S. (1977) 'Composition of economic constraints in child nutrition: impact from maternal incomes and employment in low income households', PhD Dissertation, Cornell University.

Kumari, R. (1989) *Brides are Not for Burning – Dowry Victims in India*, New Delhi: Radiant Publishers.

Lapido, P. (1981) 'Developing women's co-operatives: an experiment in rural Nigeria', in N. Nelson (ed.), *African Women and the Development Process*, London: Cass.

Lee, W. (1985) 'Women's groups in Papua New Guinea: shedding the legacy of dropped scones and embroidered pillowcases', in *Community Development Journal*, 20(3).

Lehmann, D. (1982) 'After Chayanov and Lenin: new paths of agrarian capitalsim', *Journal of Development Economics*, 11: 133–61.

—— (1986) 'Two paths of agrarian capitalism, or a critique of Chayanovian Marxism', *Comparative Studies in Society and History*, 28(4).

—— (1990) *Democracy and Development in Latin America*, Philadelphia: Temple University Press.

Levy, C. (1989) 'An introduction to the gender planning methodology in the project cycle', Mimeo.

—— (1991) 'Critical issues in translating gender concerns into planning competence in the 1990s', Paper presented at Joint ACSP and AESOP International Congress, Planning Transatlantic: Global Change and Local Problems, Oxford, UK (8–12 July).

—— (n.d.) Training materials for gender planning, 1990–92.

Liaison Committee of Development NGOs to the European Communities (LCD) (1989) *The Situation of Women in European NGOs*, NGO/GA/89/5, Brussels.

Low, A. (1986) *Agricultural Development in Southern Africa: Farm-household Theory and the Food Crisis*, London: Jamey Currey.

MacEwen Scott, A. (1986) 'Industrialization, gender segregation and stratification theory', in R. Crompton and M. Mann (eds), *Gender and Stratification*, Cambridge: Polity Press.

Mackintosh, M. (1981) 'The sexual division of labour and the subordination of women', in K. Young, C. Wolkowitz and R. McCullagh (eds), *Of Marriage and the Market*, London: CSE.

McCormack, J., Walsh, M., and Nelson, C. (1986) *Women's Group Enterprises: A Study of the Structure of Opportunity on the Kenya Coast*, Boston, MA: World Education Inc.

McIntosh, M. (1979) 'The Welfare State and the needs of the dependent family', in S. Burman (ed.), *Fit Work for Women*, London: Croom Helm, pp. 152–72.

MacPherson, S. and Midgley, J. (1987) *Comparative Social Policy and the Third World*, Sussex: Wheatsheaf Books.

Machado, L. (1987) 'The problems for women-headed households in a low-income housing programme in Brazil', in C. Moser and L. Peake (eds), *Women, Human Settlements and Housing*, London: Tavistock.

Maguire, P. (1984) 'Women in development: an alternative analysis', Mimeo, Amherst, MA: Center for International Education, University of Massachusetts.

Manser, M. and Brown, M. (1980) 'Marriage and household decision-making: a bargaining analysis', *International Economic Review*, 21.

Marsden, D. (1990) 'The meaning of social development', in D. Marsden and P. Oakley, *Evaluating Social Development Projects, Development Guidelines No. 5*, Oxford: Oxfam.

Mathai, S. (1990) 'Women and new technologies – an organizing manual', in *The Tribune*, no. 34, New York: International Women's Tribune Centre.

Mayo, M. (1975) 'Community development: a radical alternative?', in R. Bailey and M. Brake, *Radical Social Work*, London: Edward Arnold.

Mazumdar, V. (1979) 'From research to policy: rural women in India', *Studies in Family Planning*, 10.

Mead, M. (1950) *Male and Female*, Harmondsworth: Penguin.

Menefee Singh, A. and Kelles-Viitanen, A. (1987) *Invisible Hands*, New Delhi: Sage.

Metha, P. (1986) 'Evaluation of the UNICEF Indian Urban Infrastructure Programme', Unpublished report, London: Development Planning Unit.

Molner, A. and Schreiber, G. (1989) 'Women and forestry operational issues, *World Bank Policy, Planning and Research Working Paper* no. 184, Washington, DC: World Bank.

Molokonne, A. (n.d.) *The Woman's Guide to the Law*, Botswana: Ministry of Home Affairs, Women's Affairs Unit.

Molyneux, M. (1981) 'Women's emancipation under socialism: a model for the Third World', *IDS Discussion Paper DP157*, Sussex: Institute of Development Studies.

—— (1985a) 'Mobilization without emancipation? women's interests, state and revolution in Nicaragua', *Feminist Studies*, 11(2).

—— (1985b) 'Family reform in socialist states', *Feminist Review*, 21: 47–65.

Moore, H. (1988) *Feminism and Anthropology*, Cambridge: Polity Press.

Moser, C.O.N. (1978) 'Informal sector or petty commodity production: dualism or dependence in urban development?', *World Development*, 6(9/10).

—— (1981) 'Surviving in the Suburbios', *Institute of Development Studies Bulletin*, 12(3).

—— (1983) 'The problem of evaluating community participation in urban development projects', in C. Moser (ed.), Evaluating community participation in urban development projects', *Development Planning Unit Working Paper* no. 14, University College, London.

—— (1984) 'The informal sector reworked: viability and vulnerability in urban development', *Regional Development Dialogue*, 5(2).

—— (1986) 'Women's needs in the urban system: training strategies in gender aware planning', in M. Schmink, J. Bruce and M. Kohn (eds), *Learning about Women and Urban Services in Latin America and the Caribbean*, New York: The Population Council.

—— (1987a) 'Women, human settlements and housing: a conceptual framework for analysis and policy-making', in C.O.N. Moser and L. Peake (eds), *Women, Human Settlements and Housing*, London: Tavistock.

—— (1987b) 'Mobilization is women's work: struggles for infrastructure in Guayaquil, Ecuador', in C.O.N. Moser and L. Peake (eds), *Women, Human Settlements and Housing*, London: Tavistock.

—— (1987c) 'Are there few women leaders or is it that the majority are invisible?', unpublished paper presented at Conference on Local Leaders and Community Development and Participation, University of Cambridge.

—— (1989a) 'Gender planning in the Third World: meeting practical and strategic gender needs', *World Development*, 17(11).

—— (1989b) 'Community participation in urban projects in the Third World', *Progress in Planning*, 32, part 2.

—— (1989c) 'The social construction of dependency: comments from a Third World perspective', in M. Bulmer, J. Lewis and D. Piachaud (eds), *The Goals of Social Policy*, London: Unwin Hyman.

—— (1992a) 'Adjustment from below: low-income women, time and the triple role in Guayaquil, Ecuador', in H. Afshar and C. Dennis (eds), *Women and Adjustment Policies in the Third World*, Basingstoke: Macmillan.

—— (1992b) 'From residual welfare to compensatory measures: the changing agenda of social policy in developing countries, *Silver Jubilee Paper 6*, Institute of Development Studies, Sussex, UK.

Moser, C.O.N. and Levy, C. (1986) A theory and methodology of gender planning:

meeting women's practical and strategic gender needs, *DPU Gender and Planning Working Paper No. 11*, London: Development Planning Unit.

—— (1989) Draft Checklist for Participation of Women in Development Projects, Mimeo.

Moser, C.O.N. and Peake, L. (eds) (1987) *Women, Human Settlements and Housing*, London: Tavistock.

Muntemba, M.S. (1982) 'Women as food producers and suppliers in the twentieth century: the case of Zambia', *Development Dialogue*, 1–2.

Murray, N. (1979a) 'Socialism and feminism: women and the Cuban revolution, part 1', *Feminist Review*, 2: 57–73.

—— (1979b) ' Socialism and feminism: women and the Cuban revolution, part 2', *Feminist Review*, 3: 99–108.

Nash, J. (1953) 'Two person cooperative games', *Econometrica*, 21: 128–35.

Nash, J. and Safa, H. (eds) (1986) *Women and Change in Latin America*, South Hadley, Massachusetts: Bergin and Garvey.

Nashat, G. (ed.) (1983) *Women and Revolution in Iran*, Boulder, CO: Westview Press.

National Institute of Urban Affairs (NIUA) (1982) *Women Construction Workers: With Particular Reference to Legal Security and Social Justice*, Delhi: NIUA.

Netherlands Ministry of Foreign Affairs (1989) 'Women and agriculture', *Sector Paper Women and Development No. 1*, The Hague: Directorate General for International Cooperation, Ministry of Foreign Affairs.

Nimpuno-Parente, P. (1987) 'The struggle for shelter: women in a site and service project in Nairobi, Kenya', in C.O.N. Moser and L. Peake (eds), *Women, Human Settlements and Housing*, London: Tavistock.

NORAD (1985) *Norway's Strategy for Assistance to Women in Development*, Oslo: The Royal Norwegian Ministry of Development Cooperation.

Oakley, A. (1972) *Sex, Gender and Society*, London: Temple Smith.

Oakley, P. and Marsden, D. (1984) *Approaches to Participation in Rural Development*, Geneva: International Labour Organization.

Omvedt, G. (1986) *Women in Popular Movements: India and Thailand during the Decade of Women*, Geneva: United Nations Research Institute for Social Development.

Organization for Economic Cooperation and Development (OECD) (1983) doc. 18241 (29 Nov.), Paris: OECD.

—— Assistance Group Expert Group on Women in Development (1990) *Third Monitoring Report on the Implementation of the DAC Revised Guiding Principles on Women in Development*, Paris: OECD.

Overholt, C., Anderson, M., Cloud, K., and Austin, J. (1984) *Gender Roles in Development*, West Hartford, Connecticut: Kumarian Press.

Overseas Development Administration (n.d.) *Checklist for the Participation of Women in Development Projects*, London: ODA.

—— (1989) *Women, Development and the British Aid Programme: A Progress Report*, London: ODA.

Oxfam (n.d.) *Introductory Remarks on Gender Issues from GADU and an Outline of our Role in Oxfam*, Oxford: Oxfam.

Palmer, I. (1985) *The Impact of Agrarian Reform on Women, Women's Roles and Gender Differences in Development Cases for Planners*, West Hartford, Connecticut: Kumarian Press.

Pascall, G. (1986) *Social Policy: A Feminist Analysis*, London: Tavistock.

Paul S. (1987) 'Community participation in development projects: the World Bank experience', *World Bank Discussion Papers No. 6*, Washington, DC: World Bank.
—— (1991) 'Accountability in public service', *Policy, Research and External Affairs Working Paper WPS 614*, Washington, DC: World Bank.
Peake, L. (1991) 'The development and role of women's political organisations in Guyana', in J. Momsen (ed.), *Women and Change in the Caribbean*, Kingston, Jamaica: Ian Randle Publishers.
Phillips, A. (1983) *Hidden Hands*, London: Pluto Press.
Poats, S. and Russo, S. (1989) 'Training in WID/gender analysis in agricultural development: a review of experiences and lessons learned', *Working Paper Series No. 5*, Rome: Food and Agriculture Organization.
Pryer, J. and Crook, N. (1988) *Cities of Hunger: Urban Malnutrition in Developing Countries*, Oxford: Oxfam.
Redclift, N. and Mingione, E. (eds) (1985) *Beyond Employment: Household, Gender and Subsistence*, Oxford: Basil Blackwell.
Resources for Action (1982) *Women and Shelter in Tunisia: A Survey of the Shelter Needs of Women in Low-Income Areas*, Washington, DC: USAID Office of Housing.
Rogers, B. (1980) *The Domestication of Women*, London: Kogan Page.
Roldan, M. (1988) 'Renegotiating the marital contract: intrahousehold patterns of money allocation and women's subordination among domestic outworkers in Mexico City', in D. Dwyer and J. Bruce, *A Home Divided: Women and Income in the Third World*, Stanford, CA: Stanford University Press.
Rondinelli, D. (1981) 'Government decentralization in comparative perspective: theory and practice in developing countries', in *International Review of Administrative Sciences*, XLVII(2).
Rosenhouse, S. (1989) 'Identifying the poor: is "headship" a useful concept?', *Living Standards Measurement Study Working Paper No. 58*, Washington, DC: The World Bank.
Rubin, G. (1975) 'The traffic in women: notes on the "political economy" of sex', in R. Reiter (ed.), *Towards an Anthropology of Women*, pp. 157–210, New York: Monthly Review Press.
Sabatier, P. and Mazmanian, D. (1979) 'The conditions of effective implementation: a guide to accomplishing policy objectives', *Policy Analysis*, pp. 481–3.
Safier, M. (1990) 'Making plans and "making cities": creating, recognising, and sustaining "room for manoevre" in planning practice', Mimeo.
Sara-Lafosse, V. (1984) *Comedores Comunales: La Mujer Frente A La Crisis (Lima: Grupo de Trabajo)*, Lima: Servicios Urbanos y Mujeres de Bajos Ingresos.
Sargent, L. (1981) *Women and Revolution*, Boston: South End Press.
Schmink, M. (1982) 'Women in the urban economy in Latin America', *Population Council Working Paper No. 1*, New York: The Population Council.
—— (1984) 'The working group approach to women and urban services', Mimeo, Gainesville: Centre for Latin American Studies, University of Florida.
—— (1984) *Community Management of Waste Recycling: The Sirdo*', New York: The Population Council.
—— (1989) 'Community management of waste recycling in Mexico: the SIRDO', in A. Leonard (ed.), *Seeds: Supporting Women's Work in the Third World*, New York: The Feminist Press.
Schmitz, H. (1979) 'Factory and domestic employment in Brazil: a study of the

hammock industry and its implications for employment theory and practice', IDS Discussion Paper No. 146, Brighton: Institute of Development Studies.

—— (1982) 'Growth constraints on small-scale manufacturing in developing countries: a critical review', *World Development*, 10(6).

Schultz, T.P. (1988) 'Economic demography and development: new directions in an older field', in G. Ranis and T. P. Schultz (eds), *The State of Development Economics*, Oxford: Basil Blackwell.

Scott, A. and Roweiss, S. (1977) 'Urban planning in theory and practice: a reappraisal', *Environment and Planning*, 9: 1097–111.

Scott, J. and Tilly, L. (1982) 'Women's work and the family in nineteenth-century Europe', in E. Whitelegg *et al.* (eds), *The Changing Experience of Women*, Oxford: Martin Robertson/Open University.

Sebsted, J. (1982) *Struggle and Development among Self-Employed Women, A report for SEWA*, Washington, DC: USAID.

Sen, A. (1990) 'Gender and co-operative conflicts', in I. Tinker (ed.), *Persistent Inequalities*, Oxford: Oxford University Press.

Sharma, K., Pandey, B., and Nantiyal, K. (1985) 'The Chipko movement in the Uttarkhand Region, Uttar Pradesh, India', in S. Muntemba (ed.), *Rural Development and Women: Lessons from the Field*, Geneva: ILO/DANIDA.

Shiva, V. (1988) *Staying Alive: Women, Ecology and Development*, London: Zed.

Showstack Sassoon, A. (ed.) (1987) *Women and the State*, London: Hutchinson.

Small, C. (1989) 'From the ground up: an anthropological version of a women's development movement in Polynesia', Michigan State University, *Women in International Development Working Paper No. 182*.

Sollis, P. (1992) 'Multi-lateral agencies, NGOs and policy reform', *Development in Practice: An Oxfam Journal*, 2(3).

—— (1993) 'Poverty alleviation in El Salvador: an appraisal of the Cristiani Government's social programme', *Journal of International Development*, 5(1).

Staudt, K. (1983) 'Bureaucratic resistance to women's programs: the case of women in development', in E. Bonepath (ed.), *Women, Power and Policy*, New York: Pergamon.

—— (1990) 'Gender politics in bureaucracy: theoretical issues in comparative perspective', in K. Staudt (ed.), *Women, International Development and Politics*, Philadelphia: Temple University Press.

Stein, A. (1990) 'Critical issues in community participation in self help housing programmes: The experience of FUNDASAL', *Community Development Journal*, 25(1).

Stephenson, C. (1982) 'Feminism, pacifism, nationalism and the United Nations Decade for Women', *Women's Studies International Forum*, 5.

Stivens, M. (1987) 'Family and state in Malaysian industrialization: the case of Rembau, Negri Sembilan, Malaysia', in H. Afshar (ed.), *Women, State and Ideology*, London: Macmillan.

Streeton, P., Burki, S., Hag, M., Hicks, N. and Stewart, F. (1981) *First Things First: Meeting Basic Human Needs in Developing Countries*, Oxford: Oxford University Press for the World Bank.

Swedish International Development Authority (SIDA) Office of Women in Development (1989) Report from the SIDA Regional Seminar on Women and Development held in Colombo, Sri Lanka, 8–12 May 1989, Stockholm: SIDA.

—— (1990a) *Striking a Balance: Gender Awareness in Swedish Development Cooperation*, Stockholm: SIDA.

—— (1990b) Terms of reference for a SIDA Appraisal Mission, Stockholm: SIDA.

Thomas, M. (1979) 'The procedural planning theory of Andras Faludi', *Planning Outlook*, 22(2) 72–7.

Thomson, M. (1986) *Women of El Salvador: The Price of Freedom*, London: Zed.

Tijerino, D. (1978) *Inside the Nicaraguan Revolution*, Vancouver: New Star Books.

Tinker, I. (1976) 'The adverse impact of development on women', in I. Tinker and M. Bramson (eds), *Women and Development*, Washington, DC: Overseas Development Council.

—— (1982) *Gender Equity in Development: A Policy Perspective*, Washington, DC: Equity Policy Center.

Tinker, I. and Jaquette, J. (1987) 'UN Decade for Women: its impact and legacy', *World Development*, 15(3).

Turbitt, C. (1987) 'Planning for women beneficiaries: an analysis of regular programme and field programme planning systems', Mimeo.

United Nations (UN) (1976a) *World Plan of Action for the Implementation of the Objectives of the International Women's Year*, New York: United Nations.

—— (1976b) *Report of the World Conference of the International Women's Year, Mexico City, 19 June–2 July, 1975*, Sales No. E. 76. IV.1, New York: United Nations.

—— (1986) 'Forward Looking Strategies for the Advancement of Women', *Report of the World Conference to Review and Appraise the Achievements of the UN Decade for Women: Equality, Development and Peace*, A/Conf. 116/28/Rev.1, New York: United Nations.

United Nations Asian and Pacific Centre for Women and Development (UNAPCWD) (1979) *Feminist Ideologies and Structures in the First Half of the Decade for Women*, Report from the Bangkok Workshop, Kuala Lumpur: UNAPDC.

United Nations Centre for Human Settlements (UNCHS) (1984) *Community Participation in the Execution of Low Income Housing Projects*, Nairobi: UNCHS-HABITAT.

—— (1986) *The Role of Women in the Execution of Low-Income Housing Projects*, Nairobi: UNCHS-HABITAT.

United Nations Centre for Social Development and Humanitarian Affairs (UNCSDHA) (1987) *Women 2000*, Vienna: UNCHSDA Branch for the Advancement of Women.

—— (1989) *Women 2000 No. 3*, Vienna: UNCSDHA, Branch for the Advancement of Women.

United Nations Institute for Training and Research for the Advancement of Women (INSTRAW) (1988) *Modular Training Package for Women in Development*, Dominican Republic: INSTRAW.

United Nations International Children's Emergency Fund (UNICEF) (1985) 'Policy review; UNICEF response to women's concerns', E/ICEF/1985/L.1, New York: UNICEF.

—— (1987) *Urban Examples: Street Food Trade*, UE14, New York: UNICEF.

—— (n.d.) *The Invisible Adjustment: Poor Women and the Economic Crisis*, Santiago: UNICEF Americas and the Caribbean Regional Office.

United Nations Research Institute for Social Development (UNRISD) (1979) *Inquiry into Participation – a Research Approach*, Geneva: UNRISD.

United States Agency for International Development (USAID) (1978) *Report on Women in Development*, Washington, DC: USAID Office of Women in Development.

— (1982) *A.I.D. Policy Paper: Women in Development*, Washington, DC: USAID Bureau for Program and Policy Coordination.

— (1986a) *Gender Issues in Basic Education and Vocational Training*, Washington, DC: USAID.

— (1986b) *Gender Issues in Small Scale Enterprises*, Washington, DC: USAID.

— (1987a) *Gender Issues in Agriculture*, Washington, DC: USAID.

— (1987b) *Gender Issues in Natural Resources Utilization*, Washington, DC: USAID.

— Office of Women in Development (1988) *The Gender Information Framework: Gender Considerations in Design*, Executive Summary, Washington, DC: USAID.

— (n.d.) *The Gender Information Framework Pocket Guide*, Washington, DC: USAID.

Urdang, S. (1989) *And Still They Dance: Women, War and the Struggle for Change in Mozambique*, London: Earthscan.

Vance, I. (1987) 'More than bricks and mortar: women's participation in self-help housing in Managua, Nicaragua', in C.O.N. Moser and L. Peake (eds), *Women, Human Settlements and Housing*, London: Tavistock.

Walby, S. (1990) *Theorizing Patriarchy*, Oxford: Basil Blackwell.

Weeda, M. (1987) 'The role of social planning and participation in the refugee context: a question of integration or marginalization?', MSc in Social Planning in Developing Countries dissertation, London School of Economics.

White, K., Otero, M., Lycette, M., and Buvinic, M. (1986) 'Integrating women into development programs: a guide for implementation for Latin America and the Caribbean', Washington, DC: International Center for Research on Women.

Whitehead, A. (1979) 'Some preliminary notes on the subordination of women', *Institute of Development Studies Bulletin*, 10(3).

— (1981) '"I'm hungry, Mum": the politics of domestic budgeting', in K. Young *et al.* (eds), *Of Marriage and the Market*, London: CSE.

— (1984a) 'Men and women, kinship and property: some general issues', in R. Hirschon (ed.), *Women and Property, Women as Property*, London: Croom Helm.

— (1984b) 'Beyond the household? Gender and kinship-based resource allocation in Ghanaian domestic economy', Paper presented at workshop on Conceptualizing the Household: Theoretical, Methodological and Conceptual Issues, Howard Institute for Development.

— (1984c) 'Women's solidarity – and divisions among women', *Institute of Development Studies Bulletin*, 15(4).

— (1990) 'Women and rural production systems', Mimeo.

Wilensky, H. and Lebeaux, C. (1965) *Industrial Society and Social Welfare*, New York: Free Press.

Williams, S. (1983) *Women in Development: Paper Prepared for Oxfam*, Oxford: Oxfam.

Wipper, A. (1975) 'The Madaleo y Wanawake movement: some paradoxes and contradictions', *African Studies Review*, 14 (3): 99–120.

World Bank (1979) *Recognising the 'Invisible' Women in Development: The World Bank Experience*, Washington, DC: World Bank.

— (1980) *Women in Development*, Washington, DC: World Bank.

— (1984) *World Development Report 1984*, Washington, DC: World Bank.

— Women in Development Division (1990) *The World Bank Initiative for Women in Development, A Progress Report*, Washington, DC: World Bank.

World Commission on Environment and Development (WCED) (1987) *Our Common Future: The Bruntland Report*, Bruntland Report.

World Food Programme (1987) *General Guidelines*, Rome: FAO.

Yoon, S-Y. (1985) 'Women and collective self-reliance: South Korea's New Community Movement', in S. Muntemba (ed.), *Rural Development and Women: Lessons from the Field*, Geneva: ILO/DANIDA.

Young, K. (1990) 'Household resource management: the final distribution of benefits', Mimeo.

Young, K. and Moser, C. (eds) (1981) 'Women and the informal sector', *Institute of Development Studies Bulletin*, 12(3).

Young, K., Wolkowitz, C., and McCullagh, R. (eds) (1981) *Of Marriage and the Market*, London: CSE Books.

Yudelman, S. (1987) *Hopeful Openings*, West Hartford, Connecticut: Kumarian Press.

— (1990) 'The Inter-American Foundation and gender issues: a feminist view', in K. Staudt (ed.), *Women, International Development and Politics*, Philadelphia: Temple University Press.

Yuval-Davis, N. (1987) 'Front and rear: the sexual division of labour in the Israeli Army', in H. Afshar (ed.), *Women, State and Ideology*, London: Macmillan.

Name index

Subject index

abortion 41, 46
anti-poverty: as a policy approach 8, 55–8, 62, 66–9, 70, 74, 79, 93, 140, 150–1, 201, 202, 231

Bahamas 119, 120
Bangladesh 50–1, 128
Barbados 210
'basic needs strategy' 67, 68
Belize 4, 119, 120
bodies: women's control over 39, 44, 61–2, 76, 206
Botswana 54
Brazil 53
Bruntland Report (1987) 192

Canadian International Development Agency (CIDA) 127, 132, 137, 145, 158, 161, 162–3, 166, 167, 168, 169, 175
capitalism 28–9; and control of women 42–4
Caribbean countries: and institutionalization of gender planning 108, 119–22, 125, 126
Centro de Orientacion de la Mujer (COMO) 207–8
checklists: use of 4, 142, 146, 155–6, 169, 185
childbearing 27, 28, 29–31 39, 63, 101; see also reproductive role
childcare 39, 40, 43, 69, 75, 95, 142, 152, 154, 160, 167, 196, 206, 208; and father's role 53, 75; planning for 53; see also reproductive role
childrearing 27, 29–31, 48, 60, 63,

101; see also reproductive role
China 128; and population policy 45
Christian Aid: and gender planning 109, 115–18, 153; and training 175, 187; Women's Group 115–18, 127
civil society 1, 10, 90, 190, 192, 205, 207–8, 211
Colombia 197
colonialism: and women 63, 74–5, 208
Committee for the Elimination of Discrimination against Women (CEDAW) 143–4
Commonwealth Plan of Action (1987) 134
community managing: and women 27, 34–6, 40, 43, 71, 73, 91–4, 95, 100, 103, 168, 176, 196, 230
community participation: and NGOs 194–5, 206
community politics: and men 28, 35, 230; see also community managing
cost-benefit analysis: and gender 165
credit: and women 3, 39, 50–1, 52, 54, 66, 69, 99, 196
Cuba 76

data collection: and bias 97–8
debt crisis 69–70
decision-making 1, 5, 7, 104; and the household 15, 18–27, 90, 94, 97, 176, 213; and women 168, 206
Denmark 132
Development Alternatives with Women for a New Era (DAWN) 61–2, 75–6, 78, 192; and classification of women's organizations 198–203